"When it comes to explanations of hum ....e and theology often find themselves talking past each other, so it may be difficult to imagine a conversation with evolutionary biology about the profoundly theological notion of holiness. Matthew Hill not only imagines it but exemplifies it, working deftly with sociobiology and Wesleyan theology in a way that brings the two into a fruitful interaction focused on divine grace working within the restraints of creation. We are indebted to Hill for this fine display of science-faith dialogue and robust emphasis on the centrality of the church and its practices for Christian formation."

Joel B. Green, Fuller Theological Seminary

"At a time when scientific creationists and intelligent design theorists remain mired in nineteenth-century disputes about evolution, Matthew Nelson Hill's *Evolution and Holiness* moves the conversation about Darwin's controversial theory into the twenty-first century. Lamenting that 'not many theologians have examined the evolutionary data for relevance to "loving thy neighbor,"' Hill develops powerful and deeply meaningful connections between John Wesley's doctrine of Christian holiness and scientifically informed understandings of our evolved human nature. The result is a tour de force of creative theological exploration that deserves to be widely read."

Karl Giberson, scholar in residence, Stonehill College

"So many Christian books on evolution are purely defensive. Here Matthew Hill has both critique and constructive dialogue with cutting-edge science, showing how theology—and Wesleyan theology in particular—can both contribute to and learn from science in the exciting pursuit to be fully human."

David Wilkinson, principal, St. John's College, Durham University

"Theology and science have valuable insights to offer each other in understanding the origins and development of human moral and religious sentiments. Matthew Hill makes an important contribution by clearly summarizing some of the recent science on the evolution of human behavior, critiquing the materialistic philosophical extensions that are sometimes added to the science and then showing how theologically inspired practices from the Wesleyan tradition synergize with recent scientific work."

Loren Haarsma, associate professor of physics, Calvin College

"'Having trouble living the holy life? You just need to try harder!' Unfortunately, many Christians hear this message. 'Just try harder,' however, ignores the powerful role our bodies—including our genes—and the body of Christ—Christian community—must play in following Jesus' command to be holy. In this book, Matthew Nelson Hill explores the sociobiological roots of human behavior, including the constraints we all face. Along the way, Hill helps us understand altruism and generosity in ways that make sense scientifically, theologically and experientially. He argues that loving communities and their practices stand the best chance in helping us walk the highway of holiness."

Thomas Jay Oord, author of *The Uncontrolling Love of God*

"John Wesley insisted that the most compelling evidence for (1) the integrity of human choice and (2) the possibility of authentic love of God and neighbor was the life of a Christian saint, but he also recognized the value—yea, the necessity—of contesting scientific accounts of human nature and action that appeared to undercut these convictions. Matthew Hill's engagement with sociobiology is an insightful continuation of this apologetic task, defending the possibility of and offering wisdom toward the nurturing of Christian saints in our day."

Randy L. Maddox, William Kellon Quick Professor of Wesleyan and Methodist Studies, Duke University Divinity School

"What a good book! Hill's study moderates the best kind of dialogue between faith and science in which a fluent assessment of the discipline of sociobiology (in his case) interacts with a faithful understanding of John Wesley's pivotal doctrine of Christian perfection (holiness) to produce a deepened understanding of the gains of science and the practice of faith. This book exemplifies, then, a way forward in the mostly messy slugfest between faith and science that typifies this dialogue in the secular academy and evangelical church. I highly recommend Hill's programmatic discourse to faculty and clergy alike."

Robert W. Wall, Seattle Pacific University and Seminary

EVOLUTION AND HOLINESS

SOCIOBIOLOGY, ALTRUISM AND THE QUEST FOR WESLEYAN PERFECTION

MATTHEW NELSON HILL

FOREWORD BY DARREL R. FALK

IVP Academic

An imprint of InterVarsity Press
Downers Grove, Illinois

InterVarsity Press
P.O. Box 1400, Downers Grove, IL 60515-1426
ivpress.com
email@ivpress.com

InterVarsity Press® is the book-publishing division of InterVarsity Christian Fellowship/USA®, a movement of students and faculty active on campus at hundreds of universities, colleges and schools of nursing in the United States of America, and a member movement of the International Fellowship of Evangelical Students. For information about local and regional activities, visit intervarsity.org.

Scripture quotations, unless otherwise noted, are from the New Revised Standard Version of the Bible, copyright 1989 by the Division of Christian Education of the National Council of the Churches of Christ in the USA. Used by permission. All rights reserved.

While any stories in this book are true, some names and identifying information may have been changed to protect the privacy of individuals.

Cover design: Cindy Kiple
Interior design: Beth McGill
Images: Andrzej Wojcicki/Science Photo Library/Glow Images

ISBN 978-0-8308-3907-0 (print)
ISBN 978-0-8308-9900-5 (digital)

Printed in the United States of America ∞

Library of Congress Cataloging-in-Publication Data
Names: Hill, Matthew Nelson, 1981-
Title: Evolution and holiness : sociobiology, altruism, and the quest for
   Wesleyan perfection / Matthew Nelson Hill.
Description: Downers Grove : InterVarsity Press, 2016. | Series: Strategic
   initiatives in evangelical theology | Includes bibliographical references
   and index.
Identifiers: LCCN 2015040192 (print) | LCCN 2015041468 (ebook) | ISBN
   9780830839070 (pbk. : alk. paper) | ISBN 9780830899005 (eBook)
Subjects: LCSH: Holiness—Methodist Church. | Perfection—Religious
   aspects—Methodist Church. | Evolution (Biology)—Religious
   aspects—Methodist Church.
Classification: LCC BX8349.H64 H55 2016 (print) | LCC BX8349.H64 (ebook) |
   DDC 241/.047—dc23
LC record available at http://lccn.loc.gov/2015040192

| P | 23 | 22 | 21 | 20 | 19 | 18 | 17 | 16 | 15 | 14 | 13 | 12 | 11 | 10 | 9 | 8 | 7 | 6 | 5 | 4 | 3 | 2 | 1 |
| Y | 35 | 34 | 33 | 32 | 31 | 30 | 29 | 28 | 27 | 26 | 25 | 24 | 23 | 22 | 21 | 20 | 19 | 18 | 17 | 16 |

To Helene with all my love

CONTENTS

◆

Foreword by Darrel R. Falk   9

Acknowledgments   13

1 INTRODUCTION   17

  1.1 The Aim of This Study   17

  1.2 Brief Summary of Main Chapters   21

  1.3 What Is at Stake?   30

  1.4 Brief Clarifications Before Main Chapters   31

2 SOCIOBIOLOGICAL EXPLANATIONS OF ALTRUISM   37

  2.1 An Introduction to Biological Altruism   37

  2.2 Sociobiological Altruism: From Darwin to Dawkins   55

  2.3 Moving Forward   62

**3 ALTRUISM AND THE EXPLANATORY
LIMITATIONS OF EVOLUTION**   63

  3.1 The Environment and Its Influence on Human Behavior   64

  3.2 Problematic Language   72

  3.3 Reductionism and Its Relationship to the Explanation
of Altruism   76

  3.4 A Reductionist-Driven False Opposition Between
Philosophy/Theology and Sociobiology   89

  3.5 Conclusion   102

**4 OVERCOMING GENETIC AND ENVIRONMENTAL
CONSTRAINTS ON ALTRUISM**   106

  4.1 The Determined Human Person?   108

  4.2 The Human Person as Influenced but Not Determined   117

  4.3 Humans Are Genuinely Free and Consequently Responsible   120

  4.4 Conclusion   135

5   **WESLEYAN HOLINESS AGAINST
     A BACKDROP OF EVOLUTION**   137

    5.1 The Quest for Holiness   139

    5.2 Genetic Selfishness and Its Implications for Wesleyan Ethics   158

    5.3 Conclusion   169

6   **HOW WESLEY NURTURED ALTRUISM
     DESPITE BIOLOGICAL CONSTRAINTS**   172

    6.1 Wesley's Structure and Organization   175

    6.2 How Wesley Understood and Nurtured Altruism
         by Way of Holiness   192

    6.3 Environmental Constraints That Temper
         Biological Constraints   196

    6.4 Conclusion   198

7   **A LIFESTYLE OF HOLINESS**   201

    7.1 Brief Summary of Main Chapters   204

    7.2 Holiness Outside the Wesleyan Community   207

    7.3 Practical Implications and Further Explorations   210

    7.4 Conclusion   215

Appendix 1: Excerpt from "Principles of a Methodist"
            on the Topic of Christian Perfection   218

Appendix 2: Excerpt from "A Plain Account of
            Christian Perfection"   221

Appendix 3: Rules of the Band Societies—
            Drawn Up Dec. 25, 1738   224

Bibliography   227

General Index   243

FOREWORD

Darrel R. Falk

◆

This is a unique time for Christian theology. Biologists have demonstrated with near certainty that humans were created through the evolutionary process. The extensive fossil data is now in hand; we can see the physical changes that took place as the hominin body gradually became increasingly like our own bodies—step by step—over the past five million years. The archaeological data demonstrates in a remarkably unambiguous fashion the temporally well-defined development of increasingly sophisticated tools and the growing ability of hominins to modify raw materials to meet their needs. Genetic analysis of humans and other primates allows us to trace the coming and going of DNA elements that leave "watermarks" in our genomes telling us a coherent story of our common ancestry with the great apes. Clearly we have been created through the evolutionary process.

Even though the data supporting evolution is solid, this is no indication that the broader ramifications of the evolutionary process are settled. The fact is that a cadre of highly visible evolutionary biologists have yielded to the temptation to move beyond the confines of their discipline, venturing into realms defined more by personal philosophy than real scientific data. Equipped with little more than Sunday school knowledge of the Christian faith, they, with their evolution research journals in hand, purport to have scientifically demonstrated that Christianity is null and void—a mythical remnant of ancient humankind's baseless quest for meaning. In the meantime, influential evangelical leaders have attempted to show that it is science that is baseless. As they set out to dismantle the discipline of biology, they try to pull it up by its evolutionary roots and proceed to declare it to be null and void despite 150 years of all-encompassing data to the contrary. Frequently, without ever having taken a course in evolutionary biology, they have tried to develop a set of arguments to fill in the void that they think exists. So the two sides, one equipped with a childlike knowledge of Christianity, the other with little knowledge of evolutionary biology, have been busy filling in voids that likely exist only in their imaginations. In the meantime, new pathways into the theological richness of exploring God's creative processes are being ignored.

This book is a wonderful first step in initiating change. It examines a particular theological concept—the 250-year-old Wesleyan doctrine of holiness—and shows how its understanding can be enhanced and better appreciated through concepts that emerge from evolutionary biology. In actual fact, we are moving into what is potentially a very exciting time for both evolutionary biology and Christian theology. Although the basic framework laid out by Charles Darwin is clearly correct, it is now apparent that there is much more to the story. As its details have been emerging, not only are they consistent with core tenets of Christian theology, they can profoundly enrich it as well.

Natural selection as characterized by Darwin still lies at the core of our knowledge about evolution: (1) organisms *do* vary in the traits they possess; (2) the variation *does* frequently have a hereditary component; (3) some of that variation *does* result in more offspring; and (4) certain characteristics *do* come to predominate in a population over time because of the increased reproductive capacity they provide. The ramifications of this algorithm are huge. Self-centeredness, for example, is a significant outcome of the natural selection process. If your cat is hungry it is likely not willing to share a just-caught mouse with other cats, and your dog, no matter how good-tempered its breed, will likely growl if approached by another dog while it is chewing on the meat of a juicy bone.

In human evolution though, a different component started to kick in. As hominins transitioned from the trees to the lifestyle of small hunter-gatherer bands in the savannahs, traits that facilitated group dynamics began to really blossom. Characteristics like a cooperative spirit, a caring disposition, even self-sacrificing love emerged and spread. Tasks were divided up for the good of the group. As mothers more quickly weaned their young, becoming pregnant again sooner, extended family members like grandmothers, young aunts and older siblings helped care for children. This extended family became even more significant as preadult years expanded to almost two decades. Humans learned to communicate with one another through language, and they began to think in abstract terms envisaging the ramifications of their actions well out into the future. Never before had anything like this been seen—not in a billion years of multicellular life. Individuals who possessed these traits for fitting well into the group were, on average, much more successful than those who didn't, so cooperative communities flourished like never before in the animal world.

Still, even today, in part because of our ancient prehuman roots in self-centeredness, we frequently find ourselves looking out for "number one." We, like Eve listening to the voice of the serpent long ago, still hear that whisper, "Take the fruit, take the fruit." Indeed, we resonate with the words the apostle Paul penned so long ago in Romans 7:21-23: "So I find it to be a law that when I want to do what is good, evil lies close at hand. For I delight in the law of God in my inmost self, but I see in my members another law at war with the law of my mind, making me captive to the law of sin that dwells in my members." We have a divided spirit, but by grace we are called to something more, just as it also by grace through the power of the Spirit that we're enabled to respond.

Our genes have been honed through a process that includes over two million years of living together in communities, functioning as cooperative small groups, that are almost like a single body with each part taking on its own specific role. The more we know about our early heritage, the more we can understand why Paul tells us that life is most fulfilling as we live together functioning as a single body. Indeed we are called to *be* a single body—the body of Christ.

So this book by Matthew Nelson Hill, which I have found so inspiring to read, exhibits the fruitfulness that emerges as we explore the connection between Christianity and evolutionary biology. It embraces science, even as it shuns the illegitimacy of scientism. It is among the first to take the work of a highly influential figure in Protestant Christian thought—John Wesley in this case—and then demonstrate how his thinking, indeed his life's work, is enhanced as it is examined through knowledge gained from the study of human evolution. John Wesley was greatly interested in and highly knowledgeable about the scientific thinking of his time. Had he been able to fast-forward through the nineteenth and twentieth centuries into the twenty-first, this book, given Wesley's fascination with science, would have thrilled his soul. Hill shows that Wesley's bands and classes provided an accountability that resembled the small hunter-gatherer groups of perhaps thirty or so individuals, and it was in those ancient groups in which our genetic proclivity for cooperation and even love was nurtured for tens of thousands of generations.

Hill's excellent critique of the overextending reach of new atheists and of sociobiology is timely and extremely important. He shows why these approaches have been academically naive and why despite their distracting rhetoric, the two millennia of the still-unfolding Christian tradition is left largely intact regardless of the premature announcement of its death.

There is a broader theological dimension to this book that also must not be overlooked. Although Hill focuses largely on one particular aspect of the Christian tradition (holiness), he demonstrates that the approach he has taken can be adapted to evolutionarily informed studies of holiness within other theological traditions as well. Indeed, as he points out, this type of analysis is readily transferrable to theology in general.

In summary, this is a book that breaks new ground and ought to serve as a model that will be of great value to scientists, pastors, theologians and perhaps especially, laypersons. What an inspiration it has been to read the book and what a joy it is to look forward to seeing how it will stimulate a change in focus in the evolutionary biology/Christian faith discussion. No longer do we Christians need to attempt to dismantle biology; we are into a new day when we can critically examine the discipline and, given that scientists are just exploring God's truth anyway, we can celebrate the theological and philosophical beauty that emerges.

ACKNOWLEDGMENTS

◆

At the end of any significant sacrifice there are inevitable feelings of relief and dependency. So it is for me with this book. Relief speaks for itself but dependency might warrant some explaining. Through the writing process I have often felt a sense of reliance on others: guidance, personal mentoring, emotional support and the like. Yet despite this stretching process, I have realized that if I desire to be a person of sacrifice—especially for the greater good of the church and the academy—I will have life-long dependencies. If writing this book has taught me anything, I have realized that I cannot do life on my own. This reliance is both the joy and burden of living in community and my realization that we are all contingent on each other. For these reasons, I desire to thank the following individuals who have helped me through this journey.

First, I would like to thank Professor Robert Song. His guidance and direction was beyond what I could have asked for. He introduced me to a complex field with patience and grace. With every conversation we had, I walked away truly amazed at his unique perspective and intelligence. I am also forever grateful to Professor Song for changing the way I think and who I am.

Next, I would be remiss if I did not acknowledge my friends and colleagues who graciously read early drafts of this book. I want to especially thank Ken Brewer and Robert Moore-Jumonville who encouraged me to present and publish my research through various venues and critiqued my chapters on John Wesley's theology. Darrell Moore's Wesley Study

Group has also been immensely helpful, and I appreciate his kindness and mentoring. I am also thankful to various editors who read through early portions of this project: my sister-in-law Diana Welkener, who labored with me in those early stages, as well as Marie Slisher and Anna Tabone, who helped me in a pinch. I also want to thank Bryan Zinn for helping me with some research for the last chapter. I am significantly indebted to my friend Jack Baker, who tirelessly worked through this entire book. Jack is truly one of the most gifted people I know, for he has the unique ability not only to make my writing better, but to teach and guide me in the writing process while encouraging me to relax and continue on. I also want to thank Spring Arbor University for giving me a reduced teaching load for two years while I dedicated significant time to this manuscript. This reduction enabled my schedule to be much lighter and freed me up to finish well. I am also grateful for IVP—and David Congdon in particular—for both believing in this project and for their sustained support. Furthermore, I want to thank the Department of Theology at Spring Arbor University, as well as Alexander Angelov, Robbie Bolton, Brent Cline, Brian Lugioyo and Eric Slisher, who were constant sources of encouragement through the writing process. Additionally I would like to thank Kyle Poag, who stayed with me when I had a "dark night of the soul" moment and nearly gave up and walked away.

Last of all, and most significantly, I want to thank my children (Connor, Anna, Lucas and Eva), friends and family, and particularly my parents, Nelson and Karen Hill, who gave me grace throughout this journey. During the last several years I have neglected conversations, did not keep in touch, and sacrificed significant time and attention in order to finish this book. Despite all of this, they have stuck by my side and chose to understand instead of giving up on our relationship.

Yet far above all, I want to thank my wonderful wife, Helene, for her patience and love over the last few years. She has not only read and critiqued my whole project, but has also shown me what true altruism and holiness looks like, and I am grateful she allows me to share part of her life. I tear up every time I think of her kindness and only hope to be half the person she is. This book is wholly dedicated to her.

The interface between science and religion is, in a certain sense, a no-man's land. No specialized science is competent here, nor does classical theology or academic philosophy really own this territory. This is an interdisciplinary zone where inquirers come from many fields. But this is a land where we increasingly must live.

HOLMES ROLSTON III
SCIENCE AND RELIGION: A CRITICAL SURVEY

INTRODUCTION

◆

1.1 THE AIM OF THIS STUDY

A man lives in a rural part of the Ontario Province in Canada in a community of low socioeconomic status.[1] Everyone around him drinks excessively, and so does he; the difference between this man and his peers, however, is that he happens to be a First Nations Canadian compared to his European-descended friends. Unfortunately for this man, he has a genetic disadvantage that his friends do not have: he is lacking an enzyme that quickly breaks down alcohol, making him more predisposed to alcoholism than non–First Nations people.[2] Is he less morally responsible for his actions than his friends? Are there ways he might overcome the biological roulette that led to his disease? Questions like this are never easy to answer. Yet genetic predispositions like this are quite common. Perhaps the broader and more interesting question concerns how moral behavior is constrained by our biological makeup. If one can locate these influences, it might be possible to understand the environmental conditions that contribute to moral behavior.

---

[1]Although this is a fictional character, the theme is all too real. Living near a rural American reservation for several years, I witnessed this unfortunate scenario. Nearly 11 percent of First Nations people struggle with alcoholism (this is 3–4 percent higher than other ethnic groups). My suspicion is that despite unparalleled socioeconomic pressures (which are very real and I wish not to belittle them), First Nations people have to struggle through their own biological makeup in order to get past this terrible disease. I will go into this subject (and anecdote) further in chapter four. For some statistics concerning alcoholism and First Nations people, see T. Kue Young, *The Health of Native Americans: Toward a Biocultural Epidemiology* (New York: Oxford University Press, 1994).

[2]Certain First Nations ethnic groups lack key enzymes that break down alcohol efficiently (after consumption). This gives them a genetic predisposition to become alcoholics. See ibid., 210-11.

In recent years, new research in evolutionary biology and sociobi-ology has made the above scenario more intelligible, leading to the knowledge that humans possess numerous behavioral traits that di-rectly link to genetics. Yet, despite much innovative research in socio-biology—a field that applies evolutionary theory to social behavior—theological ethics is still coming to terms with what this new knowledge means for how we understand moral behavior. Therefore, this book finds itself at the intersection of theological ethics and the sciences and seeks, in part, to address questions such as these: (1) How much does sociobiology fully explain moral traits such as human altruism? (2) If genetic explanations do not fully explain human altruism, what role should we give to environmental explana-tions and free will? (3) How do genetic explanations of altruism relate to theological accounts of human goodness?[3] To move toward an-swers for these questions, I will be reading them through the lens of Wesleyan theology and ethics, offering a unique perspective within the interface of sociobiology and ethics. To propose such a reading, however, the following concerns must be considered: Can Wesleyan bands and classes provide the environmental conditions within which people may develop holiness, beyond their genetic proclivities? If so, how are we to understand Wesleyan holiness against a back-ground of evolutionary biology? Consequently, it is the central aim of this book to explore the significance of human evolutionary theory for Wesleyan holiness.

---

[3]This is a question Neil Messer specifically asks: "If we are to answer the question, 'What does it *mean* to love our neighbours as ourselves?,' by attending to the distinctive sources of Christian faith, we will not simply say, 'Altruism.'" Emphasis his. See Neil Messer, *Selfish Genes and Christian Ethics: Theological and Ethical Reflections on Evolutionary Biology* (London: SCM, 2007), 114. He continues by saying that Christians should mean "much more" than *altruism* "when asking themselves about love's meaning" and should refrain from just translating "neighbour-love into altruism." See ibid., 115. He continues, "In short, when a theological ethic that conceptualizes its task as 'understanding the Word of God as the command of God' encounters a discussion of evolution and ethics frequently obsessed with altruism, it will not only say that there is much more to Christian love of neighbour than altruism, it will also put neighbour-love itself firmly in its place: as an important part of a Christian moral vision, to be sure, but by no means all that there is to be said about the life that we are called to live in response to the love of God made known in Jesus Christ." See ibid., 128. This is a point I speak to later in the book. I believe there is a way to use the word *altruism* that can move the discussion forward without getting caught up in its definition.

For the sake of clarity, the reader should note that I purposely use a Wesleyan understanding of holiness throughout this book for two primary reasons. First, John Wesley's view of holiness was particularly stringent. While Keswickian, Reformed and even Eastern Orthodox notions of sanctification and holiness have their own interactions with human evolution (addressed later in the book), Wesleyan holiness is by far the most rigorous and extreme. I do not mention this caveat to imply any negative connotations attached to either Wesley's concept of holiness or other formulations from various parts of Christianity. Instead, I use Wesley's holiness as a lens because, if this demanding concept of holiness can connect to human evolution, then so too can the others. Second, it would simply be beyond the scope of this book to fully address all notions of holiness in every tradition. By focusing on Wesley's view, while still bringing in concepts from Reformed, Orthodox and other traditions, I hope to connect the disparate fields of holiness and sociobiology in a cogent and orderly way.

Through the process of connecting Wesleyan holiness to sociobiology, there are several other important issues on which I hope to elaborate. I will establish the connection between the sociobiological understanding of human nature, which places human behavior somewhere on a spectrum between egoism and altruism, and accountability groups that have the potential to engender generosity and altruism. I will also articulate how John Wesley, in particular, uniquely approached communicating and transmitting his social ethic. One of his primary modes was the organized bands that held members answerable to a high standard. In regard to the social context of the time, the small groups formed by Wesley, called bands and classes, encouraged people to live out a more altruistic social ethic through accountability.

What is more, from the rise of sociobiology as a self-conscious enterprise, the assumption of sociobiologists has been to explain behavior by the "selfishness" of genes, which gave rise to the problem of altruism. Altruism has been a puzzling phenomenon within sociobiology and has created problems for those who solely appeal to genetic explanations.[4]

---

[4]As Celia Deane-Drummond notes, altruism has been one of the primary "problems" for sociobiology and the subject of much debate within the field of sociobiology. See Celia

Yet, even within the field of biology, recent studies by Frans de Waal, among others, suggest that their biological "make up" puts humans more to the center of the selfish/selfless spectrum. The ability for change along the spectrum makes it possible for intentional community—structured by environmental constraints on behavior—to shift humans who are genetically inclined to both selfish and altruistic inclinations to favor either the former or latter. This type of intentional community was at the heart of the early Methodist movement. When individuals dwell within community, they learn traits and qualities that are characteristic of altruism, as can be observed in Wesley's bands and classes, which were powerful vehicles for the communal lifestyle of altruism toward the poor.[5] To such a degree, sociobiology confirms what Wesley knew to be true of human nature: humans who dwell within a system of constraints motivated by Christian perfection can cultivate altruistic behavior.

Still, within the context of these highly structured groups there are left open numerous theological and ethical puzzles that need solving. If certain individuals who are cultivating a lifestyle of holiness are more or less inclined to selfish behavior, an adequate theological explanation is necessary in order to account for these genetic differences and how they relate to the Christian call to move toward holiness. One should not separate humans into dualistic categories of body and spirit such that attention focuses only on sociobiological traits, which are inherently material. Instead, I will develop a fuller account of how Wesleyan groups function with the dual concepts of genetic/environmental constraints and ideas of Christian perfection. Although it may seem a contradiction, these concepts can form a cohesive and holistic account of the Christian person.

To be sure, the relationship between morality and holiness can sometimes seem blurry. While I recognize the differences between acts of altruism, morality and the spiritual process of becoming holy, I do not hold the position that these are unrelated activities. In much the same way as the book of James discusses the connection between works and

Deane-Drummond, *The Ethics of Nature*, New Dimensions to Religious Ethics (Malden, MA: Blackwell, 2004), 148.
[5]The kind of intentional community is evaluated depending on how richly the moral character and activity is engendered. See Stanley Hauerwas, *A Community of Character: Toward a Constructive Christian Social Ethic* (Notre Dame, IN: University of Notre Dame Press, 1981), 34.

faith, altruism and morality can be linked with holiness. As I will expound upon this throughout the book, one cannot become holy without seeing the fruit of altruism and moral development along the way. Similarly, those practicing altruism and positive moral behavior are in some sense heading in the direction of holiness. As I will discuss later, this process does not happen without the intervention of God. We can see an analogy of this from Wesley himself: when struggling with his faith, his friend Peter Böhler charged him to "preach faith till you have it and then because you have it, you will preach faith."[6] In a similar way, while practicing altruism is not tantamount to practicing holiness, they are never very far from one another.

1.2 BRIEF SUMMARY OF MAIN CHAPTERS

I intend to begin with a sociobiological look at altruism and conclude with how Wesleyan ethics might express its own account of altruistic behavior as a theological concept of Christian perfection.

The intention of chapter two, "Sociobiological Explanations of Altruism," is to address the main sociobiological narrative that altruism is a "problem" that needs explanation and solving. Within sociobiology in particular, this behavior has always been considered problematic because it seems to go against the idea of individual selection. How can something that, by definition, reduces individual gene fitness end up being a behavior trait?[7] To answer this question, and to set the tone for the following chapters, I explain sociobiological theories of altruism while showing the possibility of biological unselfishness—drawing upon

---

[6]John Wesley, *John Wesley: A Representative Collection of His Writings*, ed. Albert C. Outler (New York: Oxford University Press, 1964), 17. In *After Virtue*, Alasdair MacIntyre expresses a similar notion by using virtue language: "The immediate outcome of the exercise of a virtue is a choice which issues in right action: 'It is the correctness of the end of the purposive choice of which virtue is the cause' (l228a1, Kenny's translation, Kenny 1978) wrote Aristotle in the *Eudemian Ethics*. It does not of course follow that in the absence of the relevant virtue a right action may not be done. To understand why, consider Aristotle's answer to the question: what would someone be like who lacked to some large degree an adequate training in the virtues of character? In part this would depend on his natural traits and talents; some individuals have an inherited natural disposition to do on occasion what a particular virtue requires." Alasdair C. MacIntyre, *After Virtue: A Study in Moral Theory*, 3rd ed. (Notre Dame, IN: University of Notre Dame Press, 2007), 149.

[7]See Celia Deane-Drummond, *Christ and Evolution: Wonder and Wisdom* (Minneapolis: Fortress, 2009), 61, 160.

concepts such as kin and reciprocal altruism as well as game theory. Lastly, I investigate the contemporary conversation and the implications of such a discussion within sociobiology.

In chapter three I will discuss how sociobiological explanations of altruistic human action, especially on the basis of genetic evolution alone, are not fully satisfying. I will establish in this third chapter that, although biological explanations account for some of our understanding of altruism, they do not define the phenomena of altruism in its totality. There are also serious environmental influences that impact human action. I will also explain how habitual dimensions factor into altruistic action. Here, any action by an individual creates new behavioral patterns and norms. In other words, what biologists decipher from our genetic past does not prescribe what our future action *will* necessarily be, or what our moral behavior *should* be. There is much in the study of human action that we do not understand, even for those in the community of sociobiologists.[8]

This chapter has three major critiques of sociobiological explanations of altruism. The first criticism addresses the role of culture and its influence on learned human behavior, showing that we are not merely the products of our genes. The second critique discusses sociobiological invocations of problematic language when explaining altruism, which exposes numerous inconsistencies. The third criticism revolves around the inability of sociobiology to explain altruistic behavior without resorting to reductionism. The main issue with such oversimplification is that sociobiologists do not acknowledge the *whole* human person.[9] The last major section in this chapter builds off the former criticisms and shows how a false opposition is developed by sociobiologists between philosophy/theology and sociobiology. This false opposition causes sociobiologists to moralize outside the bounds of science as well as commit a naturalistic fallacy.

---

[8]Richard C. Lewontin, *Biology as Ideology: The Doctrine of DNA* (New York: HarperPerennial, 1992), 3. Lewontin says, "Science is a social institution about which there is a great deal of misunderstanding, even among those who are part of it."

[9]Another problem with reductionism is that it ends up giving biology explanatory priority. For more on this, see Robert Song, "The Human Genome Project as Soteriological Project," in *Brave New World? Theology, Ethics, and the Human Genome*, ed. Celia Deane-Drummond (London: T&T Clark International, 2003), 165.

One important obstacle to understanding the proper relation between Christian ethics and human evolution lies in the inappropriate forms of reductionism presumed by sociobiology and evolutionary psychology.[10] Simon Conway Morris argues that sociobiology is "not so much wrong, as seriously incomplete" when it comes to explanations for altruistic behavior.[11] Accounting for genes does not suffice for a total explanation.[12] Stephen Pope speaks to this with clarity:

> Even on the micro-level, it makes no sense to assume a genetic determinism according to which genes by themselves somehow cause behavior. Genes never function as isolated cases of behavior but, as Rose emphasizes, rather as essential components of complex networks. Behavior, moreover, reflects the influence of a multitude of genes (they are "polygenic"). Genes play an important role in the cluster of causes that lie behind behavior, but are not *"the"* cause of behavior.[13]

What Pope describes here is one of the fundamental errors of reductionist thinking. To be fair, most sociobiologists do not believe traits are the result of single genes. When they talk of a "gene for altruism" or a "gene for egoism" it can sometimes be done for linguistic convenience. Still, although sociobiologists like Dawkins are not gene "fatalists"— Dawkins specifically acknowledges that genetic causes and environmental causes are in principle no different from each other[14]—the line is often considerably blurred. This ambiguity is not isolated. Many concepts elicit opposing views that are paradoxically held by the same

---

[10]Stephen J. Pope, *Human Evolution and Christian Ethics*, New Studies in Christian Ethics (Cambridge: Cambridge University Press, 2007), 56.

[11]Simon Conway Morris, *The Crucible of Creation: The Burgess Shale and the Rise of Animals* (Oxford: Oxford University Press, 1998), 8-9.

[12]Pope, *Human Evolution and Christian Ethics*, 58.

[13]Ibid., 163. In the same section, Pope prefaces this quotation by saying, "Sociobiologists do not think that human behavior is under the 'control' of one gene or a collection of genes, nor do they hold that human traits are rigidly predetermined by genetic inheritance. Genes cannot function by themselves, and there is broad agreement that behavior is influenced by a constant interplay of learning and culture with biological predispositions and potentials rather than caused by rigidly determined behavior traits. Genes are simply chemically inert units of DNA. Placed within the appropriate cellular environment, genes direct the synthesis of proteins and these proteins in turn can generate significant effects. The expression of genes is influenced by the physical and chemical condition of the cellular environment in which they exist."

[14]Richard Dawkins, *The Extended Phenotype: The Gene as the Unit of Selection* (Oxford: Freeman, 1982), 2.

sociobiologist. For instance, there is a complicated dance that Dawkins in particular performs when he wants to call humans "merely animals" yet at the same time "set apart" from animals. Early in *The Selfish Gene*, Dawkins claims that humans are distinct from other animals:

> Be warned that if you wish, as I do, to build a society in which individuals cooperate generously and unselfishly towards a common good, you can expect little help from biological nature. Let us try to *teach* generosity and altruism, because we are born selfish. Let us understand what our own selfish genes are up to, because we may then at least have the chance to upset their designs, something that no other species has ever aspired to.[15]

One can see how, on the one hand, Dawkins wants to claim humans as animals that need to be *taught* altruism. Yet, on the other hand, humans are the only animals that can "upset" their designs. This troubled logic is inconsistent at best, and will be something I heavily critique in this chapter. Recent research is discrediting blanket statements by those, like Dawkins, who believe humans are born selfish and have to be taught how to be altruistic.[16] Primatologist Frans de Waal, in particular, has numerous studies that display how primates (some of whom are our closest nonhuman ancestors) have a biological bent toward kindness, empathy and altruism.[17] It should be clearly stated, however, that de Waal only provides qualified optimism because there are a whole host of less admirable traits that can be found among our primate ancestors[18]—a concept that I pursue throughout my critique in this chapter.[19]

---

[15]Richard Dawkins, *The Selfish Gene* (New York: Oxford University Press, 1976), 3. Emphasis his. The "selfish gene" is one of the most influential catchphrases of our times that has influenced popular understandings of biology. See also Celia Deane-Drummond, Bronislaw Szerszynski and Robin Grove-White, *Re-ordering Nature: Theology, Society, and the New Genetics* (London: T&T Clark, 2003), 123.

[16]Dawkins, *The Selfish Gene*, 3.

[17]Frans de Waal has numerous works relating to this. One prominent one in particular is Frans de Waal, *The Age of Empathy: Nature's Lessons for a Kinder Society* (New York: Harmony Books, 2009). Another good biological account that serves as an alternative to Dawkins is Kenneth M. Weiss and Anne V. Buchanan, *The Mermaid's Tale: Four Billion Years of Cooperation in the Making of Living Things* (Cambridge, MA: Harvard University Press, 2009).

[18]Messer, *Selfish Genes and Christian Ethics*, 38.

[19]Furthermore, the intersection of human evolution (and biology in general) and Christian ethical inquiries focused on altruism tend to draw the debate to reductionism. Reductionism has been the plague of the dialogue between both human evolutionists and Christian ethicists; it oversimplifies the argument and halts discussion. In a sense, both sides have sometimes crafted

In chapter four, I will discuss the biological constraints on human behavior, both genetic and environmental, and how these constraints impact human freedom and responsibility.[20] One cannot simply ask humans to "be moral," or to "be altruistic." Different people reside in different locations on the spectrum of biological and environmental constraints. I aim to demonstrate that although altruistic behavior is significantly influenced by biology, it does not mean that all altruistic actions are solely limited to neurobiological processes. Humans have the ability (unique amongst animals)[21] to overcome influences on their behavior. To do this, I will first look at some specific ways in which biology constrains human altruism. Some of these factors include the mind/body connection associated with neuroscience and genetic determinism. There are interesting parallels between altruistic action and sexual reproduction, and I will provide evidence for the similarities between these two phenomena. Next, I will show how individuals can overcome such behavioral constraints, such as defying one's genes and overcoming environmental barriers. Last, I will argue that by embracing a better understanding of freedom and responsibility, humans have the capacity to make moral decisions, including altruistic decisions, with authentic freedom.[22]

---

a straw man in order to ignore the real consequences of the claims of others. Wilson, in particular, tends toward a reductionist view of altruism, claiming that kin and reciprocal altruism account for all altruistic actions. See Edward O. Wilson, *On Human Nature* (Cambridge, MA: Harvard University Press, 1978), 167. For more reading on this, see Neil Messer's critique of Daniel Dennett's reductionist attempts to "fix" Wilson's own arguments. See Messer, *Selfish Genes and Christian Ethics*, 33.

[20]Although this chapter is not meant to be an apology for sociology against sociobiology, I do agree with Erik Erikson's idea that "it is meaningless to speak of a human child as if it were an animal in the process of domestication; or of his instincts as set patterns encroached upon or molded by the autocratic environment. Man's 'inborn instincts' are drive fragments to be assembled, given meaning, and organized during a prolonged childhood by methods of child training and schooling which vary from culture to culture and are determined by tradition." See Erik H. Erikson, *Childhood and Society*, 2nd ed. (New York: Norton, 1963), 95-96. I think much could be said of adults as well. I do not believe adults to be static and unchangeable creatures.

[21]As noted earlier, this is a point that is brought up by Dennett and Dawkins.

[22]Stephen Pope articulates this concept well by saying that "the Christian belief that each person is addressed by biblical injunctions, free to choose to obey or to disobey moral norms, and directly accountable to God generates a strong emphasis on moral agency and responsibility. The natural-law tradition gives special attention to the intellectual capacity of moral agents to comprehend the basic requirements of the moral law. The scholastics held, in other words, that the norms of the natural law were fundamentally intelligible and believed that people generally are

These environmental and genetic restrictions necessarily impact human freedom and, subsequently, responsibility.[23] However, because humans are not wholly subjugated to these influences—having the capability to overcome such constraints—it is within their power to practice genuine altruism. Nonetheless, not all individuals start at the same location on the selfish/selfless spectrum. Those who have stronger genetic drives toward selfishness will have a more difficult time overcoming those urges in order to act altruistically. Others with less biological complications, or those with environmental influences that support altruistic behavior, are more able to make altruistic decisions.

Again, it is important to note that genes alone cannot explain the totality of human behavior.[24] Daniel Dennett, for instance, is correct in saying that humans are somehow different than other animals in their capacity to live freely.[25] Yet human freedom, stemming from biological causes, is constrained by both the genetic and the environmental history of the individual.[26] Those influences contain the parameters of freedom

---

aware of the basic goods of human life." See Pope, *Human Evolution and Christian Ethics*, 280.

[23]There is a distinction between biological and cultural evolution, stemming from the ability of environments (via human culture in particular) to transmit human beliefs about matters of values to anyone who could understand their language and were persuaded by their position. In environmental evolution, those who inherit these beliefs and practices do not need to be genetically related to the sources from whom the beliefs originated. One can pass on acquired characteristics like a new set of ethical beliefs, mores or practices to one's offspring, making cultural evolution more Lamarckian rather than Darwinian. See Philip Clayton, "Biology and Purpose: Altruism, Morality, and Human Nature in Evolutionary Perspective," in *Evolution and Ethics: Human Morality in Biological and Religious Perspective*, ed. Philip Clayton and Jeffrey Schloss (Grand Rapids: Eerdmans, 2004), 319. For instance, one can hold a false belief that does not fare well in the selection process without dying in the process in cultural evolution. Biological mistakes, however, are much more severely punished by nature, resulting in natural selection. This cultural evolution is at a faster pace than natural selection through genetic variation and selective retention by the environment. Again, see the helpful chapter of Clayton, ibid.

[24]See Pope, *Human Evolution and Christian Ethics*, 163. Philip Clayton also discusses some practical reasons why socialization plays an enormous role in forming individual selves. In large extent, parents are the "socializers" of children. Yet the age at which humans are accepted into society as fully functioning adults is getting later. Whereas it used to occur in early teenage years, now college serves as a sort of "psycho-social moratorium," to use a phrase from Erik Erikson, and has shifted that date into one's early twenties. See Clayton, "Biology and Purpose," 329.

[25]Daniel C. Dennett, "Animal Consciousness: What Matters and Why," *Social Research* 62, no. 3 (1995).

[26]I should mention that evolutionists tend to "protect" the idea of free will by emphasizing the fact that genes and phenotype will always have an interaction with environment and culture; and behavior will always be conditioned by this interaction. See Pope, *Human Evolution and Christian Ethics*, 166.

but do not fully hamper genuine choice. Each individual differs in where they dwell on the continuum of constraints. These constraints can be located in both one's family of origin and the communities one belongs to.[27]

Chapter five is titled "Wesleyan Holiness Against a Backdrop of Evolution." John Wesley created an environment for the early Methodists by which holiness might be better actualized. Yet significant issues between the concept of Wesleyan holiness and genetic predispositions remain. Although the intersection of Wesleyan ethics and sociobiology has not been explored at great length, even within current research the Wesleyan perspective on Christian perfection is conspicuously missing as compatible with sociobiological explanations of human behavior. This explanatory problem is particularly troubling when it concerns moral behaviors, including altruism.

For instance, as we have seen, sociobiologists have found evolutionary links to prosocial behavior. These discoveries might appear to put the idea of Christian perfection in jeopardy. It would seem that an individual does not need to be concerned about experiencing Christian perfection if she or he is bound by genetics. Or, it might make more sense to simply cultivate inherited behavioral traits. Likewise, if genetic understandings of *Homo sapiens* convey that humans are predisposed to certain moral behaviors, either being more selfish or more selfless than others, does the notion of Christian perfection mean that humans no longer have to worry about their genetic history because they can "spiritually overcome" such heritages? Then one might wonder if "perfected" Christians somehow use free will to "trump" genetics.

Within the intersection of sociobiology and Wesleyan ethics rest other important explorations I will address in this chapter. I will show that explaining human selfishness as a mere "defect of the will" is reductionist at best; instead it is a defect of the *person*. By this I will show that Wesleyan ethics helps explain how the holistic person impacts selfish and selfless behavior. One cannot have only care for either the "spirit" or "body" in some kind of reductionist dualism. Furthermore, when an individual works within her or his community, especially a community such as John

---

27Ibid., 183.

Wesley's highly structured groups, the individual can develop the kind of character necessary to be able to overcome genetic constraints.

To make the connection between sociobiology and Wesleyan ethics—specifically the concept of Christian perfection—I will first introduce Wesley's theological underpinnings, including the concepts of original sin and prevenient grace. Next, I will use the concept of *theosis* to elucidate Christian perfection in light of sociobiological discoveries. Last, I will discuss genetic selfishness and its implications for Wesleyan ethics—positing a new theory about how one might overcome genetic constraints and subsequently reach Christian perfection.[28]

In chapter six, "How Wesley Nurtured Altruism Despite Biological Constraints," I will discuss how, given that the biological human condition rests somewhere between total selfishness and altruism, there are some environmental constraints that end up pushing individuals closer to altruism.[29] John Wesley found a way to work within biological constraints, while utilizing environmental constraints, to encourage people to be more selfless through his bands and classes. It is obviously anachronistic to say that Wesley knew about the biological human condition. Instead, he placed people in groups for both the practical reasons of organization and the theological reasons of engendering holiness. This environment is what I will call Wesley's "world of constraints." Through this, he nurtured group members' biological proclivities toward altruism and mitigated egoistic tendencies.

To establish this claim, I will first look at the organization of the early Methodist groups. I also feel it is important to understand some termi-

---

[28]This chapter also engages the spiritual dimension of the early movement. It is important for the reader to remember that Wesley saw the movement as having a spiritual source that included religious experiences. He saw the Holy Spirit as taking an active role in the eighteenth-century revival and formulated a theology of Christian perfection to accompany this movement and influence his groups.

[29]As stated earlier, sociobiology gives a poor account (at least not a total account) of why altruism and morality persist in human behavior. And, as Messer notes, instead of reductionist accounts a better theological ethic proves more coherent. See Messer, *Selfish Genes and Christian Ethics*, 248. John Wesley does this exact thing, albeit unknowingly. Messer claims that the "ongoing difficulties that have been associated with the concept of altruism become easy to understand when we realize that it is a secularized, and thereby truncated and distorted, version of the Christian concept of ἀγάπη." See ibid. I would go one step further, following the path of Wesley, and claim that holiness, properly speaking to encompass inward and outward holiness, would incorporate a full understanding of altruism and ἀγάπη.

nology as well as the culture of the early Methodists before discussing the reasons behind such groups. I will walk through the specific account-ability structure that included a host of environmental constraints put forth by Wesley. It is also important to account for Wesley's theological understanding of holiness, which was parsed into *inward* and *outward* holiness. As we will see, theology was practical for Wesley, and his sermons became tools for spiritual formation.[30] Understanding this will demonstrate to us the foundation with which Wesley was working and help us to know how his groups nurtured altruism. While his core theo-logical concepts included a distinct view of the doctrine of original sin that was coupled with prevenient grace,[31] a direct connection can be made to current sociobiological discoveries of human nature not being totally selfish but also bearing altruistic behavior and potential.[32]

Wesley engaged the biological human condition in order to promote the holistic altruism that was at the heart of his drive toward holiness. When individuals dwelled within these intentional communities, they exhibited the distinguishing virtues of selflessness and altruism. This chapter finishes with some practical questions. Primarily, how does someone living in the era of Frans de Waal, understanding that the human condition is not totally depraved by egoism, follow Wesley's ex-ample and engender Christian holiness?

Besides being a catalyst for revival in England and America, Wesley articulated, through practical service to social needs, what it meant to be a Christian living in the world. For Wesley, living among the poor was part of working out one's salvation. In this way, being socially active was

---

[30]John Wesley, *Sermons 1–33*, ed. Albert C. Outler, The Bicentennial Edition of the Works of John Wesley (Nashville: Abingdon Press, 1984), xiii.

[31]In response to how these ideas lived out in reality (as well as other soteriological questions), Wesley states in his sermon 43, "The Scripture Way of Salvation," "The salvation which is here spoken of is not what is frequently understood by that word, the going to heaven, eternal hap-piness. . . . It is not a blessing which lies on the other side of death. . . . It is a present thing. . . . [It] might be extended to the entire work of God from the first dawning of grace in the soul till it is consummated in glory." John Wesley, *Sermons 34–70*, ed. Albert C. Oulter, The Bicentennial Edition of the Works of John Wesley (Nashville: Abingdon Press, 1985), 156. The major focus for salvation, for Wesley, was transforming the here and now. For reference to this, please see Darrell Moore, "Classical Wesleyanism" (unpublished paper, 2011). I am indebted to conversa-tions with Moore and to this essay for a number of ideas in this chapter.

[32]This will again touch on studies by Frans de Waal.

a key component for John Wesley's ministry and discipleship model.[33] Wesley seemed to intuitively understand the human condition to be moveable on the selfish/selfless spectrum. With the creation of his world of constraints, his intention was to move people toward inward holiness that resulted in outward holiness. This environment helped nudge Wesley's followers ever closer to what it means to live a life lived sacrificially for others. As Alasdair MacIntyre says, moral training frames our responses to our body and allows us to gain distance from our response to the "good of our animal nature."[34]

1.3 WHAT IS AT STAKE?

Human action is often (maybe almost always) generated by mixed motives.[35] This fact cannot easily be ignored by those who may want to reduce every act of altruism to a "single egoistic motivation" of a selfish gene or environmental constraint.[36] One example of this is that humans often display extraordinary sacrificial, and consequently altruistic, behavior that reduces reproductive success without compensatory reciprocation or benefit to kin.[37] One could look to countless anecdotes to see individuals sacrifice for strangers, individuals practicing celibacy to better serve others and so on. Consequently, it is possible to have actions that are freely made to the detriment of the agent.

Nevertheless, all human action will necessarily be found within the context of community. The character of morality (or even the natural

---

[33] Paul Wesley Chilcote, *Recapturing the Wesleys' Vision: An Introduction to the Faith of John and Charles Wesley* (Downers Grove, IL: InterVarsity Press, 2004), 50.

[34] Alasdair C. MacIntyre, *Dependent Rational Animals: Why Human Beings Need the Virtues*, The Paul Carus Lecture Series (Chicago: Open Court, 1999), 49.

[35] As Thomas Jay Oord notes in his *Defining Love*, even the definitions of intention and motivation have multiple interpretations and are often blurry. See Oord, *Defining Love: A Philosophical, Scientific, and Theological Engagement* (Grand Rapids: Brazos Press, 2010), 15-18.

[36] See Pope, *Human Evolution and Christian Ethics*, 224. Here, Pope continues, "Even admitting the prevalence of selfish predispositions does not mean that we have to acquiesce in a theory of universal 'psychological egoism,' the claim that people always pursue what they think is in their self-interests, or that we must ignore the presence of genuine altruism when and where it is found." The tenor of Messer's book echoes this claim as well; see *Selfish Genes and Christian Ethics*.

[37] Jeffrey P. Schloss, "Emerging Accounts of Altruism: 'Love Creation's Final Law'?," in *Altruism and Altruistic Love: Science, Philosophy, and Religion in Dialogue*, ed. Stephen G. Post et al. (New York: Oxford University Press, 2002), 221.

law) constrains the person to moral standards that encourage the well-being of the person and her or his community: it is not right to be sexually unfaithful, murderous, a thief and so on.[38] Thus, proper Christian community "does not level off the sharp edge of individuality but demonstrates the fathomless goodwill that defines Christian love."[39] What is at stake is the preservation of Christian communities to recognize the responsibility to follow what John Wesley did: capitalize on a biological phenomenon given to us by human evolution. To fail to do this would be disastrous for communities and would neglect the call to become a holy people; this would happen through succumbing to the temptation to be selfish while lacking inner transformation.[40] As Neil Messer says,

> The Church, for all that it has too often been part of the problem rather than part of the solution, is called to live as a community that embodies and bears witness to this transformation, in ways that also have the potential to challenge and change the wider society within which it is located.[41]

If the church can muster up an intentional community that can take the biological state of humanity and encourage it to be more holy (which would lead toward a fuller expression of altruism compared to reductionist sociobiological definitions), then what is at stake is not only the future of the church but also the fine tuning of human evolution.

1.4 BRIEF CLARIFICATIONS BEFORE MAIN CHAPTERS

The question of the sociobiological explanation of altruism is experiencing renewed attention and has considerable impact on contemporary Christian ethics. In 2007, two pieces of work in the area of human evolution and Christian ethics influenced how an understanding of evolution may impact the church and ethics. Stephen Pope, a Roman Catholic, published a book called *Human Evolution and Christian Ethics*,[42]

---

[38]Pope, *Human Evolution and Christian Ethics*, 292.

[39]Mildred Bangs Wynkoop, *A Theology of Love: The Dynamic of Wesleyanism* (Kansas City, MO: Beacon Hill Press, 1972), 29.

[40]Christian Smith, Michael O. Emerson and Patricia Snell, *Passing the Plate: Why American Christians Don't Give Away More Money* (Oxford: Oxford University Press, 2008), 125.

[41]Messer, *Selfish Genes and Christian Ethics*, 205.

[42]Pope has written more than this text alone. Yet, in my estimation, and for the purposes of this study, this work is the most appropriate. See Pope, *Human Evolution and Christian Ethics*. See

in which he discusses the origins of morality and engages with evolutionary theory on its own terms. The second work, *Selfish Genes and Christian Ethics*,[43] written by Neil Messer—a Reformed Protestant—seeks to find a cogent theory of morality that goes beyond evolutionary theory alone while dealing justly with it. There are, of course, many other works that have recently influenced this field.[44] But for the purposes of my argument, these two works are both exigent and pertinent—thus, I will refer to them frequently.

Since I find their works to be convincing, I will not be arguing with either Pope's or Messer's positions; instead, I wish to build off their arguments by looking into the normative aspect of how the church can foster moral actions given biological predispositions. I will develop a framework to convey how human action is not destined to be entirely selfish, and demonstrate what can be achieved if certain conditions on individuals can be met. In building this framework, I will be drawing on much of Frans de Waal's research with primates and present the connection between our ancestors and our natural prosocial tendencies. Of course, as might be expected, I will be dealing with Richard Dawkins's idea of selfish genes.[45] Although I have numerous disagreements with his and

---

also Stephen J. Pope, *The Evolution of Altruism and the Ordering of Love*, Moral Traditions & Moral Arguments (Washington, DC: Georgetown University Press, 1994).

[43]See Messer, *Selfish Genes and Christian Ethics*. Again, I will assume Neil Messer's position, and subsequently attempt to move the conversation forward while building upon his work. He says that "an account of human being as moral being, developed on the basis of a theological understanding that the world and humans have been created 'very good' by God, is well able to assimilate the proposal that aspects of moral experience emerged as a result of evolutionary process that gave rise to our species. Such a theological account can also take from human evolutionary history a useful reminder that human persons, whose personal identities are constituted by the history of their relationships with God and one another in the world, are also physically embodied beings to whom some possibilities are open and others not." See ibid., 246-47.

[44]Another work, above others, is Thomas Jay Oord's 2010 book, *Defining Love*. I will also refer to this regularly as it deals with scientifically influenced theological/philosophical positions regarding love and altruism.

[45]It must quickly be mentioned and put aside that Dawkins's objection to Christianity is intellectually questionable. He presupposes that the proper understanding of all life comes from the natural sciences and that the most satisfactory explanation of human behavior is provided in evolutionary terms. This claim leads to the assumption that all references to the transcendent are illusions that must be rejected by logical and rational people. See Pope, *Human Evolution and Christian Ethics*, 13. Thus, Dawkins claims to strictly limit his conversation to the biological, claiming that Darwin alone provides a solution of how unordered atoms could group themselves into more complex patterns until they manufacture people. See Dawkins, *The Selfish Gene*, 13.

other sociobiologists' reductionistic readings of the human condition, I agree that we have many naturally selfish tendencies. Thus, I will show how humans are caught somewhere toward the center of the selfish/ selfless spectrum, arguing, therefore, that it takes a *certain kind* of community to encourage[46] individuals to be more altruistic.

Messer also suggests, and I think rightly so, that a Christian account of the created world should locate humans as *part of* God's good creation, and that right moral action should be defined, especially by Christians, as "going 'with the grain' of God's good purposes in creation."[47] He goes on to claim that this understanding can better address how moral claims call for an authentic concern for others and why such claims matter.[48] Being in agreement with his articulation, I do not intend to introduce a new theological paradigm. I do, however, acknowledge what Messer says: "Even if we understand our natural inclinations properly, they will not necessarily (so to say) tell us the truth about our good."[49] Instead, what I intend to do is exhibit how John Wesley worked within the restraints of natural tendencies, redirecting and reshaping them to help create an environment within which holiness was possible, and by which people were able to become more altruistic.

To make the argument of this book more cogent, it is necessary to pause and outline a key clarification. Altruism, which is the most complicated of the terms in this book, is traditionally defined as an action motivated by a concern for the welfare of at least one other person—a definition that contains four primary components:[50] (1) to have a certain

---

Yet this becomes highly problematic when he moves to deal with moral problems.

[46]Through the course of this research, I have chosen to use this word (*encourage*). Yet, do the constraints that humans put on each other "push," "nudge" or "coerce" individuals' natural inclinations? See chapter four for my response to this question.

[47]Messer, *Selfish Genes and Christian Ethics*, 248.

[48]Ibid.

[49]Ibid., 119. I agree with Messer when he states, "In contradistinction to modes of Christian apologetic engagement that tend to proceed by trimming the claims of Christian faith to fit the confines of a scientific world-view. . . . A full-blooded articulation of a particular Christian tradition, sharp corners intact and rough edges un-smoothed, can offer rich resources for a coherent and fruitful engagement with issues raised by the natural sciences." Ibid., 249.

[50]All four components are taken from Pope, *Human Evolution and Christian Ethics*, 222. Sociobiologists tend to argue that agents who act in apparently phenotypic altruistic ways are actually acting in egotistic ways because their real motivations are hidden from others or themselves. This is problematic for sociobiologists and evolutionary psychologists when they make claims

kind of psychological motivation—that is, to act on behalf of others; (2) to understand and make a judgment on the worthiness of the motivation—that is, is it truly good?; (3) to decide to act on the basis of this judgment about what is good—that is, to become the directing motive of the altruistic act; and (4) to intend to do something—that is, to pursue a certain course of action to obtain the desired good. Another clarification about altruism concerns what Pope suggests: "An act need *not* be *purely* or *exclusively* self-sacrificial or involve heroic levels of self-denial on the part of an agent to count as genuinely altruistic."[51]

Not all people share the same genetic predispositions. A nonreductionist reading of motivation holds that genetics is only one of multiple factors that influence an individual's particular motivation.[52] For instance, a person who, through a traumatic incident, developed a psychologically unhealthy disposition may have different motivations for action than someone who has lived a sheltered life. To this end, looking at altruistic action as solely stemming from biological urges does not take traumatic brain injuries and other extrinsic factors into account.

In *Selfish Genes and Christian Ethics*, Messer promotes a healthy caution toward the word *altruism*, pointing out the word's complicated history. Sociobiologists tend to use the word to imply that everything kind (or even moral) can be explained through kin selection or reciprocal altruism.[53] This, however, is not the case. As Messer points out,

> Some evolutionary explanations have a tendency to reduce "morality" to a relatively small number of behaviours and traits, exemplified by *altruism*: there is sometimes a tendency to assume that if altruism can be explained in evolutionary terms, it should also be possible to explain the other behaviours and traits that make up morality. But from a theological standpoint, how satisfactory is this as an account of what we mean by morality? Within the

---

about human motivation, while at the same time professing a stance that neglects to consider internal states. See also ibid., 221.

[51]Ibid., 224. Emphasis his. See also C. Daniel Batson, *The Altruism Question: Toward a Social Psychological Answer* (Hillsdale, NJ: L. Erlbaum, Associates, 1991).

[52]Pope, *Human Evolution and Christian Ethics*, 223.

[53]For an attempt to understand altruism as a combination of group-selected altruism and cultural evolution, see Christopher Boehm, *Hierarchy in the Forest: The Evolution of Egalitarian Behavior* (Cambridge, MA: Harvard University Press, 1999). See also P. J. Richerson and R. Boyd, "Complex Societies: The Evolutionary Origins of a Crude Superorganism," *Human Nature* 10 (1999).

framework of a Christian theological anthropology, what is meant by it, and is it the kind of thing that *could* be susceptible to a natural scientific (for example, evolutionary) explanation?[54]

Messer's cautions are helpful, and I do not wish to dispute them in this book. However, for the sake of moving the conversation forward, I will be using the word *altruism* throughout. This term forms a cohesiveness between acts of self-sacrifice and kindness, which are rooted in free will, environment or religion, and sociobiological accounts of the same phenomena. For the purposes of my argument, I will attempt to stay within the confines of Pope's definition of altruism rather than that of many sociobiologists, in particular E. O. Wilson's definition: "When a person (or animal) increases the fitness of another at the expense of his own fitness."[55] I realize and acknowledge that altruism, especially taking into account a Christian anthropology (which I unapologetically hold), is deeper than mere biological urges.

Besides not wanting to get caught in the semantics of what altruism is, moving the conversation forward adds not only to the field of sociobiology but also to the field of Christian ethics in general and Wesleyan ethics in particular. Understanding what altruism might look like in religious communities provides specific relevance to Christians because humans share prosocial tendencies with animals, and this can sometimes seem quite threatening.[56] Beyond the surface of the evolution/

---

[54]Messer, *Selfish Genes and Christian Ethics*, 84. Emphasis his. In discussing morality more broadly, Messer later articulates that "there is a great danger of circularity in the more reductionist attempts to explain morality in evolutionary terms: 'morality' is equated with those features (such as altruistic behaviour) for which plausible evolutionary explanations can be given, and it is then proclaimed that evolutionary biology can explain 'morality' without remainder. By contrast . . . human morality has its origins in the goodness of God's creation, [and] is shaped by the God-given ends and goals of human life, [which] includes a responsibility (extending well beyond enlightened self-interest) for the non-human creation, and is subject to radical questioning and re-conception in the light of the death and resurrection of Christ." Ibid., 96.

[55]Edward O. Wilson, *Sociobiology: The New Synthesis* (Cambridge, MA: Belknap Press of Harvard University Press, 1975), 117.

[56]Thomas Jay Oord shares insight into the complicated relationship between Christians and many scientists. He says, "For more than a century now, many Christians have been apprehensive about what they believe are implications of evolutionary theories. This worry has little to do with the age of the earth or the fossil record—although these matters are important. Instead, the worry has to do with the apparent continuity between humans and the rest of the animal world." Oord, "Morals, Love, and Relations in Evolutionary Theory," in Clayton and Schloss, *Evolution and Ethics*, 287. Oord is correct that Christians are less concerned about the age of the earth and

creation debates—which I do not intend to visit beyond this mention—
understanding altruism holds enormous significance for the church.[57] If
we can understand the biological constraints that have been placed on
us by nature, then we, just like John Wesley, might be able to capitalize[58]
on such a phenomenon in order to encourage the church to be a more
altruistic Christian community. This altruistic community would be assembled with a holistic approach, taking the entire person into account,
rather than focusing only on the person as an embodied spirit, or some
other iteration of dualism.

---

more concerned about somehow thinking that if we are connected to animals, then we lack the
imago Dei. For a look at how not only Christians but other religions see the science/ethics conversation, see Thomas Jay Oord, *The Altruism Reader: Selections from Writings on Love, Religion,
and Science* (West Conshohoken, PA: Templeton Foundation Press, 2008).

[57]In October 1996, Pope John Paul II said in a speech to the Pontifical Academy of Sciences that
evolution was "more than a hypothesis," which has eased some Roman Catholic tensions on the
creation/evolution debate. In *Nature as Reason*, Catholic theologian and ethicist Jean Porter also
argues that "contrary to what is commonly assumed, the scholastics' view do not conflict with
the doctrine of evolution, or more generally with the biological sciences." See Porter, *Nature as
Reason: A Thomistic Theory of the Natural Law* (Grand Rapids: Eerdmans, 2005), 51. St. Thomas
Aquinas seems to accept Aristotle's definition of a human as a "rational animal," which conveys
the essential meaning. See Craig A. Boyd, "Thomistic Natural Law and the Limits of Evolutionary Psychology," in Clayton and Schloss, *Evolution and Ethics*, 223.

[58]Although this seems like an aggressive word choice, I believe this adequately grasps the method
of John Wesley.

2

SOCIOBIOLOGICAL
EXPLANATIONS OF ALTRUISM

◆

2.1 AN INTRODUCTION TO BIOLOGICAL ALTRUISM

The quest for an answer to the "quandary of altruism" leads somewhere beyond biological or sociobiological explanations alone. These disciplines cannot, on their own, be expected to provide a sufficient basis for comprehending altruism or cooperative behavior, as I will argue in chapter three.[1] In order to show the inadequacies of sociobiological explanations alone, one ought first to review the biological account of how organisms are understood within evolutionary theory, as I intend to do in this chapter.

For some evolutionary biologists, organisms came about in order to act as hosts for "replicator genes." In a rapid changing environment, replicators—forced to compete with harsh consequences, where only those best adapted to the particular environment survive—had to find means of adapting faster than other entities less able to adapt. Cooperation in groups, then, arose because it conferred adaptive advantage.[2] Thus, as eons passed, genes in cooperative groups were

---

[1]For more on this, see Stephen J. Pope, *The Evolution of Altruism and the Ordering of Love*, Moral Traditions & Moral Arguments (Washington, DC: Georgetown University Press, 1994), 103.
[2]Cooperation is always found in a benefits-to-cost ratio, whether knowingly or not. See Martin A. Nowak, "Five Rules for the Evolution of Cooperation," in *Evolution, Games, and God: The Principle of Cooperation*, ed. Martin A. Nowak and Sarah Coakley (Cambridge, MA: Harvard University Press, 2013), 109. Also, altruism is sometimes thought to be a subset of cooperation. "The form of psychological cooperation called altruism . . . is a subset of evolutionary cooperation."

selected to sacrifice in the short-term in order to benefit their individual gene fitness (via the group) in the long-term. The payoff in the end was greater than unadulterated selfishness, and a form of altruism had begun. Yet, according to Philip Clayton, sociobiologists disagree on whether biology is sufficient to account for altruism or whether a source is required that lies completely outside the purview of evolutionary theory.[3] This dispute is especially evident when dealing with *Homo sapiens*. If replicator genes try to find the best-suited host through the natural selection process, humans are a very good expression of that "host" due to their capability to be highly adaptive in various environments. The point of contention concerning standard sociobiology is that there is no room left for an agent to be "purely" altruistic, even if it does enjoy benefits of some kind, either through reciprocal behavior or kin preferences. Genuine altruism in humans, then, gets typically examined under the microscope of the selfish gene theory.[4]

Cooperative behaviors count as altruistic only when carried out by an agent when (1) she possesses cognitive, affective, and dispositional states of sufficient complexity that they can count as motivating her actions, and (2) at least some of her actions are motivated by 'goodwill' or 'love for another.'" See Philip Clayton, "Evolution, Altruism, and God: Why the Levels of Emergent Complexity Matter," in Nowak and Coakley, *Evolution, Games, and God*, 346.

[3] Philip Clayton, "Biology and Purpose: Altruism, Morality, and Human Nature in Evolutionary Perspective," in *Evolution and Ethics: Human Morality in Biological and Religious Perspective*, ed. Philip Clayton and Jeffrey Schloss (Grand Rapids: Eerdmans, 2004), 318. Still, Dobzhansky's Dictum states that nothing in biology makes sense except in light of evolutionary theory; this is evidenced by most evolutionary anthropological efforts that are aimed at showing how cultural development is a product of evolution. See Theodosius Dobzhansky, "Nothing in Biology Makes Sense Except in the Light of Evolution," *American Biology Teacher* 35 (1973).

[4] As Daniel Dennett mentions, evolution is "famously shortsighted." See Daniel C. Dennett, *Freedom Evolves* (New York: Viking, 2003), 194. When animals evolve, they are not concerned with long-term fitness, as if the genes that compose an animal were personified and autonomous. On the contrary, animals look for short-term gains. Therefore, at first glance, evolution seems to work against the practice of altruism, because altruism typically pays off at the end of the long process of intergenerational sacrificing, not during it. Stephen Clark puts it rather succinctly: "The notion of 'the selfish gene,' as a way of expounding neoDarwinian theory, at first sight seems an obvious and clumsy metaphor. 'Being selfish' is being inclined to give one's own wishes greater weight than can be justified. Genes presumably have no wishes, and whether they would give them more weight than they should, who knows? 'Altruistic behaviour,' on the other hand, is willingly doing good to others, at some personal cost. Who knows what costs there are for genes, or what 'doing good' to them requires?" See Stephen R. L. Clark, *Biology and Christian Ethics* (Cambridge: Cambridge University Press, 2000), 129. It is unclear to what extent humans are influenced by such genetic programing; yet one thing is clear, the genes that compose a person have been chosen by natural selection because they are *functionally* beneficial.

2.1.1 *Darwinian history*

According to Darwinian history, the beginning was simple. Darwin's theory of evolution by natural selection[5] aims to show how simplicity can change into complexity over time: molecules group themselves into more complex patterns, moving toward the complexity found in *Homo sapiens*.[6] All life hails from simple beginnings, most probably in hot undersea vents where basic molecules such as amino acids and RNA are assembled. These molecules are then able to replicate and mutate to adapt to their environment, and then copy that change through multiplication. For evolution to happen, there has to be the correct blend of both mutation and replication. If replicators did not possess the ability to copy and reproduce, then the whole process is random at best and would not build and progress. Yet, having the ability to copy oneself, replicator molecules drifted in the sea, where they found stability quicker and more efficiently than others and survived the primordial chaos.[7] Due to the nature of environmental conditions (for example: limited resources in the sea), as well as the earth in general, these replicators proved fitter to survive the competition of other contending life.[8] After hundreds of millions of years, replicators stopped floating in the sea, having pushed and competed their way to more stable forms of living within "hosts." As Dawkins puts is, "Their preservation is the ultimate rationale for our existence. They have come a long way, those replicators. Now they go by the name of genes, and we are their survival machines."[9] Consequently, the story of the replicator lives on in their host organism.

Stephen J. Gould later developed an important concept called "punctuated equilibrium" that proposes that geological tempo of speciation differs radically from gradual anagenesis, or the progressive evolution of

---

[5]While Darwin did not have the comprehensive view of "units of selection" as we have today, *The Origin of Species* provides a precursor to selection units of genes, individuals and even groups. Units of selection, therefore, are the biological traits that are selected for by natural selection. Positive or otherwise beneficial traits enable better reproductive fitness for the individual or group (composed of individuals). See Charles Darwin, *The Origin of Species*, The Harvard Classics (New York: Collier, 1961), chaps. three and four.

[6]Ibid., 241. Darwin himself does not mention atoms, but his theory certainly implies that notion.

[7]Richard Dawkins, *The Selfish Gene* (New York: Oxford University Press, 1976), 16-19.

[8]Ibid., 20.

[9]Ibid., 21.

species.[10] The idea of anagenesis stems from species formation that does not subdivide from the evolutionary line of descent.[11] This differs from cladogenesis, or the process of adaptive evolution that leads to the development of a greater variety of animals or plants.[12] Here, Gould proposes the notion that certain periods in evolutionary history moved more rapidly (punctuated) than others and resulted in cladogenesis where species split in the evolutionary process.[13] Gradual anagenesis, which was the commonly held timetable for evolutionary thinking before Gould, did not account for the adaptive qualities of some organisms that pointed toward short bursts of change in the midst of much evolutionary stasis.[14] For the laity, it might not seem like a major discovery in science. Yet, this new theory is noteworthy due to the necessary impact of non-biological influences on the host that inevitably come from such short bursts of evolutionary change. In such a rapidly changing evolutionary environment, forced to compete through harsh environmental circumstances, replicators had to find means of progressing faster with greater stability in order to out-replicate their competitors.

In trying to compete for fitness and stability, cooperation became a fruitful strategy by enhancing group fitness, which ultimately increased individual fitness. For instance, if a group had members that sacrificed for the community, that group would out-select groups that did not sacrifice in such an efficient way. This kind of group selection, in turn, aided cooperative groups—and the individual organisms that compose such groups—in their evolutionary competition. It was thought, prior to the 1960s, that most adaptation to selection, where an organism struggles for survival of the fittest, was considered adaptive and a benefit to the whole group. Yet, recent reviews of the evolutionary process show that the individualistic emphasis of Charles Darwin's original theory is probably more valid.[15] It just so happens that these individuals resided *within*

---

[10]Stephen Jay Gould, *Punctuated Equilibrium* (Cambridge, MA: Belknap Press of Harvard University Press, 2007), 54.
[11]See Oxford English Dictionary, 2nd ed., s.v. "anagenesis."
[12]See Oxford English Dictionary, 2nd ed., s.v. "cladogenesis."
[13]Stephen Gould and Niles Eldredge, "Punctuated Equilibria: The Tempo and Mode of Evolution Reconsidered," *Paleobiology* 3, no. 2 (1977): 145.
[14]Gould, *Punctuated Equilibrium*, 54.
[15]Robert M. Axelrod, *The Evolution of Cooperation* (New York: Basic Books, 1984), 89. It is worth

groups; thus, the fate of both the individual organisms and the groups were intimately woven together. Accordingly, replicators in groups traded short-term sacrifice for long-term benefits, and altruism became a valuable and necessary action.

Still, it was unclear if cooperation itself could be considered, as Thomas Jay Oord calls it, "absolutely altruistic,"[16] especially if the replicators in organisms were more likely to survive if they looked after their own gene fitness. Richard Dawkins attempted to address this very issue. In his book, *The Selfish Gene*, he endeavors to elucidate the genetic drive for reproductive fitness found in nature, describing the reasoning behind what appears to be cooperative actions. In the end, he denies the likelihood of any human altruistic motives in the pure sense of the word. He contends that all behavior is based upon genetic foundations that have a long evolutionary history of survival and have determined, to a large extent, just how organisms behave. He states, "The fundamental unit of selection, and therefore of self-interest, is not the species, nor the group, nor even, strictly, the individual. It is the gene, the unit of heredity."[17] This genetic makeup affects the phenotypic behavior—or the observable characteristics when genes interact with the environment[18]—and also functions as the impetus for all action and motivation for decision making, whether conscious or unconscious. Dawkins asserts that genes function as "replicators" whose primary responsibility is to survive by whatever means necessary.[19] Accordingly, the organism merely functions

---

noting that multilevel selection theory (to be discussed at a later portion of this book) tries to incorporate both individual and group benefits of cooperation. See also Lee Cronk and Beth L. Leech, *Meeting at Grand Central: Understanding the Social and Evolutionary Roots of Cooperation* (Princeton: Princeton University Press, 2012).

[16]Thomas Jay Oord, *Defining Love: A Philosophical, Scientific, and Theological Engagement* (Grand Rapids: Brazos Press, 2010), 80.

[17]Dawkins, *The Selfish Gene*, 12. It should be noted that the genesis for these potent genes came from the idea of evolving proteins. After eons of time, proteins evolved from the primordial soup. These molecules copied themselves repeatedly and eventually various strands of deoxyribonucleic acid (DNA) developed. This DNA contains the genetic blueprint that instructs living organisms. The role of DNA was to multiply itself and instruct the organism to perform actions necessary for survival. The basic idea is that genes that survive to the next generation direct all organisms due to their adaptive ability and reproduction. See Craig A. Boyd, "Thomistic Natural Law and the Limits of Evolutionary Psychology," in Clayton and Schloss, *Evolution and Ethics*, 227.

[18]Oxford English Dictionary, 2nd ed., s.v. "phenotype."

[19]Dawkins, *The Selfish Gene*, 16.

as a vehicle for the replicators' survival. The genes that are the most fit will survive; thus, the most productive mechanism to insure the best longevity is an openness to constant adaptation. Therefore for Dawkins, if the genotype—the genetic construction of an organism—determines not only the phenotype, but also largely the organism's behavior, organisms do not possess the ability to resist the genetic urgings of the replicators, thus rendering any concept of altruism unlikely.[20] Consequently, the complicated phenomenon of altruism—and whether it is beneficial, self-destructing, or able to be "purely" practiced—became a "problem" to be solved within sociobiology.[21]

2.1.2 Kin preference and reciprocal altruism
Darwin poses a serious question about acts of altruism under the assumption that the purpose of our existence is to survive and reproduce. If this is true, one should question why an individual would risk such altruistic behavior. If there is not adequate reciprocation, the individual might suffer some kind of fitness loss. Darwin says,

It is extremely doubtful whether the offspring of the more sympathetic and

---

[20]Boyd, "Thomistic Natural Law and the Limits of Evolutionary Psychology," 227.

[21]E. O. Wilson attempts to build on a question that Darwin asked: "How can altruism, which by definition reduces personal fitness, possibly evolve by natural selection?" This is a powerful question, one that is understandably troublesome to sociobiologists. If one believes that altruism is in fact utterly selfless then it does not fit inside a Darwinian framework. Again, for the purposes of this discussion, the definition of altruism is the practice (or possibly even the belief) of concern for the well-being of others. This definition often times involves acts of selflessness whereby the agent of altruism does not receive any benefit from the recipient. Yet, even for Darwinists, it seems possible to preserve some semblance of an altruistic society because the notion that all creatures that are altruistic are bound to negatively impact gene fitness is simply a straw man argument. There are other ways of creating fitness than by means of violence, fraud and selfishness. Note Conor Cunningham, *Darwin's Pious Idea: Why the Ultra-Darwinists and Creationists Both Get It Wrong* (Grand Rapids: Eerdmans, 2010), 40. Martin Nowak has a similar take: "Evolution is based on a fierce competition between individuals and should therefore only reward selfish behavior. Every gene, every cell, and every organism should be designed to promote its own evolutionary success at the expense of its competitors. Yet we observe cooperation on many levels of biological organization. Genes cooperate in genomes. Chromosomes cooperate in eukaryotic cells. Cells cooperate in multicellular organisms. There are many examples for cooperation among animals. Humans are the champions of cooperation: from hunter-gatherer societies to nation states, cooperation is the decisive organizing principle of human society. No other life form on Earth is engaged in the same complex games of cooperation and defection. The question of how natural selection can lead to cooperative behavior has fascinated evolutionary biologists for several decades." See Nowak, "Five Rules for the Evolution of Cooperation," 99.

benevolent parents, or of those that were the most faithful to their comrades, would be reared in greater number than the children of selfish and treacherous parents of the same tribe. He who was ready to sacrifice his life, as many a savage has been, rather than betray his comrades, would often leave no offspring to inherit his noble nature. The bravest men, who were always willing to come to the front in war, and who freely risked their lives for others would on average perish in larger numbers than other men. Therefore it seems scarcely possible . . . that the number of men gifted with such virtues, or that the standard of their excellence, could be increased through Natural Selection.[22]

If this is the case, one should ask, then why sacrifice in the first place? Given what we know about the modern synthesis, where genetics is combined with evolutionary theory, it might not benefit genes one bit to be self-sacrificing or altruistic.

Michael Ruse, a philosopher of science, approaches the altruism phenomenon by arguing that kin altruism and reciprocity explain the bulk of altruistic acts. For Ruse, there is a connection between a sense of sympathy and likelihood of reciprocity. Organisms have evolved to where they may be dependent on one another through social bonds—such as flock, herd or other community—or are at least organisms that are capable of reciprocating good deeds. Still, Ruse often argues that genetic reciprocators, such as direct kin, are more influential over any nonbiological obligation (whether social, cultural, moral etc.).[23] Accordingly, "biologically, our major concern has to be towards our own kin, then to those at least in some sort of relationship to us (not necessarily a blood relationship) and only finally to complete strangers."[24] These concerns, therefore, are not out of any sort of "pure" altruistic desire, but out of the biological drive to receive payback for any amount of exertion or self-sacrifice—a process that fosters a more fertile environment for gene fitness.

---

[22]Charles Darwin, *The Descent of Man and Selection in Relation to Sex* (New York: Penguin, 2004), 155.

[23]For further reading on this concept, see Stephen Pope, "Relating Self, Others, and Sacrifice in the Ordering of Love," in *Altruism and Altruistic Love: Science, Philosophy, and Religion in Dialogue*, ed. Stephen G. Post et al. (New York: Oxford University Press, 2002), 168-81.

[24]Michael Ruse, *Taking Darwin Seriously: A Naturalistic Approach to Philosophy* (New York: Blackwell, 1986), 106. And see Michael Ruse, "Evolutionary Ethics: A Phoenix Arisen," *Zygon* 21, no. 1 (1986): 103.

It is the role of a gene, then, as many sociobiologists might attest, to function without caring about its own fitness but rather about the immortal genetic set of replicas existing in other related organisms.[25] This long-term genetic urge causes the host organism to act in altruistic ways—or at least not competitively.[26] According to biology, when it comes to gene self-preservation, genes will do anything that enhances their chances of replication. This drive for multiplication trumps even the host organism itself. What makes complex organisms a worthy host for such a powerful force is that they are able or willing to be "self-deceived" in order to attain gene fitness and perpetuation of their genotype.[27] Consequently, authors such as Richard Alexander would say, "When we speak favorably to our children about Good Samaritanism, we are telling them about a behavior that has a strong likelihood of being reproductively profitable."[28] This subversive and even subconscious notion is that the host's replicators might reap beneficial results in subsequent generations from such a behavior through creating a more fertile environment for gene fitness—that is, helping to create a world that is more amiable to the replicator's next generation.

Regarding this kind of gene fitness, there are two major theories of why organisms display altruistic behavior: kin preference, also known as kin altruism, and reciprocity. Kin preference is when an agent is inclined to assist another because of genetic relation whereas reciprocity is due to some kind of positive tradeoff that might manifest itself at a later date. These two major theories of assistance giving do not logically have to depend on firm egoistic presuppositions.[29] The term *kin* in sociobiology terminology refers solely to genetic relationships.[30]

---

[25] The reader might notice the personification language used when discussing genes. I will critique this later in the book.

[26] Axelrod, *The Evolution of Cooperation*, 89.

[27] This phenomenon is interesting in light of manipulation altruism as well. Here, the attempt is by the host to manipulate others for their (or more accurately, their genes') behalf, while simultaneously trying to detect deception to avoid becoming deceived. See R. Dawkins and J. R. Krebs, "Arms Races Between and Within Species," *Proceedings of the Royal Society of London: Series B, Biological Sciences* 205, no. 1161 (1979).

[28] Richard D. Alexander, *Darwinism and Human Affairs*, The Jessie and John Danz Lectures (Seattle: University of Washington Press, 1979), 102.

[29] Pope, *The Evolution of Altruism and the Ordering of Love*, 114.

[30] Stephen Garrard Post, *Unlimited Love: Altruism, Compassion, and Service* (Philadelphia: Templeton

2.1.2.1 *Kin altruism*

Genetics drive organisms to be inclined to favor their own kin.[31] Degrees of closeness in relation to kin also matter in regard to favor. For instance, a father baboon will have genetically more in common with his son than with his cousin. In turn, however, he would have more in common with his cousin than with a baboon in the same colony. Yet he would still have more in common with that baboon than one from a different area or genetic line. Thus, he would be more inclined to aid a baboon depending on proximity or degree of closeness.[32] In theory, a cycle in which altruism was genotypically advantageous became more and more prominent among organisms at various stages in the evolutionary process. With such communal advantages, organisms in groups such as these formed better systems of protection, developed communication (dolphin sonar, crow calls etc.) and inevitably established culture for symbolic means of social coordination.[33]

Take crows, for example, who are highly sociable and communicative. During periods of overpopulation or times when food is scarce, they tend to reduce their numbers by having fewer young.[34] As pointed out by David Lack in his *Population Studies of Birds*, many animals put the interest of kin over those who are unrelated. Almost without exception, those that do not are actually sacrificing for *family* rather than unrelated groups.[35] Likewise, wolf packs, ants, as well as crows all dwell in very

---

Foundation Press, 2003), 75. However, the notion of kin in some organism's sphere, such as primate adoption, often include those who are not genetically related and receive the same status as genetically connected members. See also Pope, *The Evolution of Altruism and the Ordering of Love*, 114.

[31]Stephen J. Pope, *Human Evolution and Christian Ethics*, New Studies in Christian Ethics (Cambridge: Cambridge University Press, 2007), 216.

[32]Evolutionary theory of kinship holds that prior to the Neolithic revolution, early hominoids lived in small communities of hunter-gatherers. Over the course of millions of years, those who cooperated and worked with each other produced more offspring and gained a reproductive advantage over their counterparts. Thus, their genetic line reproduced at a higher rate, ultimately affecting the gene frequency within the population. See Pope, *The Evolution of Altruism and the Ordering of Love*, 117.

[33]Ibid.

[34]Vero Copner Wynne-Edwards, *Animal Dispersion in Relation to Social Behaviour* (New York: Hafner, 1962).

[35]Matt Ridley, *The Origins of Virtue: Human Instincts and the Evolution of Cooperation* (New York: Viking, 1997), 176; David Lambert Lack, *Population Studies of Birds* (Oxford: Clarendon Press, 1966).

large families, not unrelated groups. Any sacrifice they endure is for kin altruistic purposes. After all, if an unrelated group reduced the number of offspring during a time of scarce food, the rogue selfish individual animal would produce more kin and in a few generations all genetic propensity toward altruism would have been evolved out of the group.[36]

These systems, as well as related genotype, contributed to group loyalty and protection within genetically related groups. One can see that more complex organisms intrinsically have both competitive and altruistic-cooperative behavior that is dependent upon what promotes reproductive fitness better at any given time. Altruism, then, is regarded as involving a complex symbolic system that ensures cooperation within the group and protects it from threats by outside groups.[37] An individual organism may even take on losses to their own well-being to enable an organism with high relatedness (especially within an animal's immediate family).[38] The evolution of the honeybee worker, which has a suicidal barbed sting, might be an example of this.[39] All of this selfless behavior promotes the replicator gene hosted within the organism and its closest kin.

William Hamilton sets himself up along the same line of thinking by introducing the connection between the selfish gene and altruism in "The Genetic Evolution of Social Behavior (I and II)," wherein he argues that altruistic behavior could be a very good strategy if one is helping others who share the same copies of genes as oneself, and is thus reproduced by proxy, so to speak.[40] This surrogate gene replication is still

---

[36]For these examples of crows and wolves see Lack, *Population Studies of Birds*. Another example of individual selection overriding group selection would be in the gender ratio of 50:50 among almost all animals, of which Matt Ridley provides an interesting example. He says that if a female rabbit had the power to alter her reproduction and produced only sons, each son having ten mates, she would have, at first, ten times the amount of grandchildren, and males would eventually take over the whole species. However, then a rabbit would come along with the ability to produce females and bring the cycle back to 50:50. Consequently, the group might change dynamics, but would ultimately be recalibrated to the individual. Ridley's account is found in Richard D. Alexander, *The Biology of Moral Systems*, Foundations of Human Behavior (Hawthorne, NY: A. de Gruyter, 1987), 102-3.

[37]Pope, *The Evolution of Altruism and the Ordering of Love*, 117.

[38]Axelrod, *The Evolution of Cooperation*, 89.

[39]W. D. Hamilton, "Altruism and Related Phenomena, Mainly in Social Insects," *Annual Review of Ecology and Systematics* 3, no. 1 (1972): 193-232.

[40]W. D. Hamilton, "The Genetical Evolution of Social Behaviour 1," *Journal of Theoretical Biology* 7, no. 1 (1964): 1-16, and "The Genetical Evolution of Social Behaviour 2," *Journal of Theoretical Biology* 7, no. 1 (1964): 17-52.

aimed at helping the individual, and against any definable altruism toward society at large. According to basic Darwinian thinking, there need not be a reason to care for society unless it positively affects gene fitness. There is even the thought that "pity" may be understood as altruistic under the concept of the selfish gene. People may act egotistical even in their pity because of both relief that the pain is not happening to them and pity because the pain on them could be on the individual as well.[41]

2.1.2.2 Reciprocal altruism and game theory

Besides being genetically disposed to be altruistic toward one's kin, organisms may be driven toward altruistic action if they have historically received reciprocation. The theory of reciprocal altruism complements the notion of kin selection with the idea that it may be genetically advantageous for those who frequently interact with each other, yet are not biologically related. As Michael Ruse puts it, reciprocal altruism has a "you scratch my back, I'll scratch yours" functionality.[42] For biological reasons, it is suggested by such proponents as Ruse that one does not "scratch a back" without the forward thinking of when their "back can be scratched" in return because such behavior would not only lead to them being taken advantage of if there is no reciprocity,[43] but also goes against their genetic hardwiring. David Axelrod, in *The Evolution of Cooperation*, calls it a "tit for tat" relationship.[44] This is a strategy that shows how cooperation based on reciprocity can thrive in a varied environment with numerous competing strategies of cooperation falling short. Cooperating with cooperators and not cooperating with "cheats" seems to be the most evolutionary stable strategy.[45] Some examples of reciprocity include

---

[41]Thomas Hobbes, *Leviathan: Or, the Matter, Forme and Power of a Commonwealth, Ecclesiasticall and Civil*, Collier Classics in the History of Thought (New York: Collier Books, 1962), 53. Today, reductive scientists might posit mirror neurons as an example of this phenomenon. See Stein Bråten, *On Being Moved: From Mirror Neurons to Empathy*, Advances in Consciousness Research (Amsterdam: John Benjamins Pub., 2007).

[42]Michael Ruse, "Evolutionary Ethics Past and Present," in Clayton and Schloss, *Evolution and Ethics*, 43.

[43]Pope, *Human Evolution and Christian Ethics*, 217-18.

[44]Axelrod, *The Evolution of Cooperation*, 54.

[45]See the mention of this axiom in classic philosophy such as Aristotle, *Nicomachean Ethics*, trans. Martin Ostwald, The Library of Liberal Arts 75 (Indianapolis: Bobbs-Merrill, 1962), 1169b16-22, 62a16-19.

the fungus and alga that compose a lichen,[46] the ants and ant-acacias where the trees feed the ants who protect the trees,[47] and the fig wasps and the fig tree where the parasitic wasps serve as the tree's sole means of pollination and seed set.[48]

E. O. Wilson takes the concept a bit further by saying that altruism in which there is no benefit to the giver is not altogether foreign. He states that altruism must be extended to both members outside the group who may be in need and who are "unable or unwilling" to reciprocate, as well as to reciprocators that benefit the giver.[49] Wilson sees this as a critical building block to social harmony—which, in a peripheral way, benefits the giving agent. This might be one reason why he categorizes acts of selflessness into his concepts of "hard-core" and "soft-core" altruism, creating a sliding scale of altruistic acts.[50]

Of course, some acts of altruism are obviously not immediately reciprocated; however, one has to keep in mind that the evolutionary idea of reciprocity reflects the idea that the rewards are, on average, greater than the costs.[51] So, in the long run, the "back will get scratched," though not necessarily by the organism that received the altruistic act. This phenomenon is called "indirect reciprocity," where "Organism A" might cooperate with "Organism B" who cooperates with "Organism C" who, in turn, cooperates with "Organism A." In a sense, we might do well to imagine that genes survived the natural selection over generations of altruistic organisms due to an overall reciprocation in the environment and better gene fitness. Reciprocity and kin preference are then best understood as complementary predispositions that evolved through natural selection over the millions of years of the evolutionary past. However, it should be noted that the reciprocity theory is not a notion of how each discrete act of reciprocity is somehow consciously or unconsciously

---

[46]Axelrod, *The Evolution of Cooperation*, 90.

[47]D. H. Janzen, "Coevolution of Mutualism Between Ants and Acacias in Central America," *Evolution* 20 (1966): 249-75.

[48]J. T. Wiebes, "A Short History of Fig Wasp Research," *Gardens Bulletin* (1976): 207-32; D. H. Janzen, "How to Be a Fig," *Annual Review of Ecology and Systematics* 10 (1979): 13-52.

[49]Edward O. Wilson, *On Human Nature*, 25th anniversary ed. (Cambridge, MA: Harvard University Press, 2004), 155-59.

[50]Pope, *The Evolution of Altruism and the Ordering of Love*, 126.

[51]Alexander, *The Biology of Moral Systems*, 160.

calculated to play to the agent's inclusive fitness advantage.[52] Consequently, over time, the genes progress further, even when each act of altruism is neither immediately nor directly reciprocated.

Game theory provides helpful insight into reciprocal altruism by explaining how ranges of organisms advance their gene fitness and by modeling how reciprocity can emerge from selfish behavior over long periods of time. Over such periods of time, organisms fare better if they act in ways that are cooperative. The genes that comprise organisms that act in altruistic ways are more likely, over time, to survive. An organism does not need a brain, per se, to function in this way. For example, bacteria are highly responsive to selected aspects of their environment, which allows them to modify their behavior in response to what other organisms around them are doing. As these behaviors are selected for, these genotypes can be inherited and passed on as a productive adaptation.[53] Yet bacteria have no memory to catalog these interactions for reciprocity's sake. Other organisms, such as intelligent primates, that are high in neural complexity on the evolutionary spectrum have more intricate game-strategy behavior than those with low neural complexity. Their rather complex memory, especially when compared to the bacteria, gives them the ability to retain information that might help them strategize with whom to reciprocate altruism and with whom to compete.[54]

Cooperation, then, among biological organisms has to meet certain criteria in order to survive in genes. Axelrod stresses that the organism has to have three major developments.[55] First, the organism has to have what he calls "robustness," which is when an organism has a certain type of strategy that can thrive in diverse environments composed of others using a variety of more or less sophisticated strategies. Second, the organism has to have the ability to resist invasion by mutant strategies (apart from altruism)—this he calls "stability." Third, the organism has to have "initial viability" where the robust and stable strategy gets a foothold in an environment that is predominantly noncooperative. It is

---

[52]Pope, *The Evolution of Altruism and the Ordering of Love*, 118.
[53]Axelrod, *The Evolution of Cooperation*, 94.
[54]Ibid.
[55]All criteria are taken from ibid., 95.

here where a "tit for tat" strategy of reciprocal altruism seems to aid the organism in achieving this goal of robust stability and viability.[56]

The "tit for tat" strategy can thrive in diverse environments, which makes it uniquely beneficial for replicator genes. "Tit for tat" is a certain kind of strategy derived from the "Prisoner's Dilemma" where two "players" pair off in a reward/punishment game. They can either cooperate or defect, but both hold reward and punishments depending on how the players, collectively and individually, choose.[57] "Tit for tat," as related to cooperation, is based on the idea that reciprocal behavior, either for good or bad (or in the case of the Prisoner's Dilemma, for cooperation or defection) will result in the player making the choice of the previous action taken by the opposing player.[58] In the case of altruism, then, action such as cooperation gets reciprocated for both the benefit of the acting agent and the receiving organism.[59]

Altruism and cooperation among the organisms can evolve when the conditions of cost, benefit and possibly (but not necessarily) relatedness yield net gains for the altruism-causing genes that reside in the related individuals.[60] Nondefecting in a single-move Prisoner's Dilemma is altruism of a kind because the agent is forgoing benefits that might have been taken by the agent. In effect, the organism has a certain amount of

---

[56]Ibid., 96. Game theory is helpful in understanding reciprocal altruism, but it also sheds some light on how sociobiology views *freedom of choice* within altruistic actions. Typically, game theory considers "accurate" calculations against "free choices." For example, if someone acts in a way that would go against their own interest, then they have made an inaccurate calculation according to what might be "best" for the situation. Likewise, an organism that goes against its own fitness is seen as acting abnormally. In the case of nonkin altruism and nonreciprocal altruism, any human acting out on such behavior would be an irregularity. Thus, as we will discuss in chapter four, altruism is only specifically free—at least in the way that sociobiologists like Wilson might suggest—in regard to its profitability for the altruistic agent. See Alexander, *The Biology of Moral Systems*, 102-3. See also Pope, *Human Evolution and Christian Ethics*, 180.

[57]Axelrod, *The Evolution of Cooperation*, 9.

[58]Ibid., 13.

[59]As a result, Daniel Dennett's word, *benselfishness*, may best define the human condition (he also uses words like *quasi-altruism* or *pseudo-altruism*). See Dennett, *Freedom Evolves*, 197. In any case, there are numerous examples of altruistic behavior that benefits both the agent and the recipient. Thus, to individuals like Dennett, true or authentic altruism is "an elusive concept," an idyllic notion that evaporates whenever someone tries to contain it. See also ibid., 194. Thomas Jay Oord, in his *Defining Love*, 80, calls this kind of altruism "absolute altruism" and rightly finds the concept to be untenable.

[60]W. D. Hamilton, "The Evolution of Altruistic Behavior," *The American Naturalist* 97, no. 896 (1963): 354.

interest in the partner's gain.[61] This type of favoritism (especially kin altruism) can be seen in the spawning relationships of sea bass. These fish have both male and female sexual organs and take turns being the "high investment partner" (laying the eggs as opposed to fertilizing with sperm).[62] Biologist Eric Fischer has suggested that inbreeding would necessarily imply relatedness in the pairs and this would promote cooperation with the need of further relatedness.[63] In the end, with this kind of cooperative strategy, the giving organism is benefited with a better environment for gene fitness.[64]

---

[61] Axelrod, *The Evolution of Cooperation*, 97. One should be aware that as generations of humans passed, being selfless merely for the sake of being selfless was unproductive, and in the long run, nonreproductive. Too much altruism posed a threat to human replicators. Under the Prisoner's Dilemma, one might wonder about defection, motivation and selflessness. It is rational to end up defecting if the other person defects, for instance, and cooperating would be a poor choice. If they do not defect, then the rewards are substantial. The question is this: If everyone knows this, what is the motivation for cooperation? To put it in evolutionary terms, if payoffs are short-term, eons of evolution would be influenced by short payoffs. Thus, since one can see that co-operation does in fact exist, it must have *some* evolutionary benefit. In other words, there are forms of purely selfless actions (or pure altruism), but these actions, at least in regard to game theory, would not be beneficial to the agent. Any sustained actions like this would be self-defeating (and thus anomalies in this theory). Consequently, many altruistic acts are highly influenced by this evolutionary process, and will have some kind of benefit to the agent or the agent's kin. See Matt Ridley, *The Red Queen: Sex and the Evolution of Human Nature* (New York: Macmillan Publishing Company, 1994), 94.

[62] Axelrod, *The Evolution of Cooperation*, 98.

[63] Eric A. Fischer, "The Relationship Between Mating System and Simultaneous Hermaphroditism in the Coral Reef Fish, Hypoplectrus Nigricans (Serranidae)," *Animal Behaviour* 28, no. 2 (1980): 620-33.

[64] In regard to altruistic behavior among humans, from a practical standpoint, the act of selflessness can provide some positive benefit to the agent, and not merely strip them of all that was sacrificed. But receiving some benefit does not necessarily make a particular action nonaltruistic. Individuals with altruistic reputations can fair better, as demonstrated with the Prisoner's Dilemma. Habitually altruistic persons can cooperate effectively in scenarios where cheating might be impossible to detect. Thus, what can be viewed as "genuine altruism" can emerge solely on the reputation the agent has built. This scenario of the Prisoner's Dilemma can be seen in more detail in Ridley, *The Origins of Virtue*, 180. This idea of genuine selfishness that potentially benefits the agent goes directly against Ayn Rand's idea of what altruism looks like practically. Rand says that "[altruism] permits no view of men except as sacrificial animals and profiteers-on-sacrifice, as victims and parasites—that it permits no concept of a benevolent co-existence among men—that it permits no concept of *justice*. . . . To rebel against so devastating an evil, one has to rebel against its basic premise. To redeem both man and morality, it is the concept of '*selfishness*' that one has to redeem." See Ayn Rand and Nathaniel Branden, *The Virtue of Selfishness: A New Concept of Egoism* (New York: New American Library, 1965), xii-xiii. Yet merely because one happens to gain any benefit from an altruistic action does not mean the integrity of the action was compromised. This is a false attribution fallacy. Robert Frank, author of *Passions Within Reason*, shows that altruists do not have to be impoverished people. Rather, experimental studies often find a positive correlation with altruists and economic status (though

One tension within the field of sociobiology is the debate over how genetic traits get transferred from generation to generation—whether kin selection (also known as "inclusive fitness theory") or group selection (also known as "multilevel selection theory") is the mode by which replicators efficiently transfer beneficial traits such as altruism. These disputes have recently increased due to the 2012 publication of E. O. Wilson's book, *The Social Conquest of Earth*.[65] This book is not without criticism; the outcry stems from a 2010 publication to the journal *Nature*, in which Wilson (along with two colleagues) supported multilevel selection theory.[66] Consequently, Wilson's 2012 book-length project met the disapproval of some evolutionary theorists who claim, owing to Wilson's endorsement of multilevel selection theory, that he misunderstands kin selection (or at least overlooks it).

The core of the debate focuses on whether or not inclusive fitness has better explanatory power than multilevel selection theory. Inclusive fitness theory is a concept that quite neatly explains why individual genes might benefit from their host's self-sacrificial or altruistic actions. Kin selection can explain reasons why an organism might act altruistically and sacrificially—even to the detriment of the altruistic agents themselves. Besides helpful explanations for insect eusociality, inclusive fitness selection also helps sociobiologists explain why human selflessness might be present.[67] On the other hand, multilevel selection theory—which has explanatory capabilities and should not be discounted—is not seen as a tight argument when compared to kin selection theory, yet is gaining popularity within the field of sociobiology.[68] Essentially, groups that cooperate together have better survival odds compared to groups that

---

this is not a causal relationship). Still, it does demonstrate that altruism itself is not a depreciating characteristic or drain on human existence (from an evolutionary perspective), nor does an act become less moral if some benefit was received. See Robert H. Frank, *Passions Within Reason: The Strategic Role of the Emotions* (New York: Norton, 1988), 91.

[65]Edward O. Wilson, *The Social Conquest of Earth* (New York: Liveright Pub. Corporation, 2012).

[66]Martin A. Nowak, Corina E. Tarnita and Edward O. Wilson, "The Evolution of Eusociality," *Nature* 466, no. 7310 (2010). This article generated dozens of negative responses, many of which claim the "descent of Wilson" and his bastardization of evolutionary theory.

[67]I am indebted to a colleague, Professor Michael Buratovich, for his dialogue about both kin selection and group selection that helped clarify some key points for me regarding this debate.

[68]The tension between kin selection and group selection theorists could come as a result of "turf wars" as well.

remain uncooperative.[69] Naturally, groups could be overly altruistic, which might lead to too many individuals sacrificing, and thus to the weakening of the group. Yet multilevel selection theory holds that there is a "goldilocks zone" where individuals within groups that are cooperative, but not overly selfless, would outperform individuals within groups where members are mostly selfish. As Elliott Sober and David Sloan Wilson report in *Unto Others*, for altruism to evolve in a multilevel selection theory, "the differential fitness of groups (the force favoring the altruists) must be strong enough to counter the differential fitness of individuals within groups (the force favoring the selfish types)."[70] Oord also states that in multilevel selection theory individuals "act altruistically toward members of their group so that the group as a whole survives and thrives."[71] In this way, the group as a whole (made up of individuals) does better; and this is why inclusive fitness theory and multilevel selection theory do not have to be incompatible and perhaps not as divorced from each other as some might assume.[72]

Darwin attempted to introduce a somewhat problematic idea of group selection whereby evolutionary pressure selects and acts upon groups instead of individuals.[73] Darwin's idea of group selection, however, differed from the multilevel selection theory of today—as Darwin

---

[69]It is important to remember that for sociobiology, a *group* is not as significant as the *individuals* who compose that group. It might be good to point out here that sociobiology is an attempt to convince people that human life is pretty much what it has to be and perhaps even ought to be. Yet underneath it lies a deeper ideology, which is the priority of the individual over the collective. Despite the name *socio*biology, we are dealing with a theory not of social causation but of *individual* causation. The characteristics of society are seen as caused by the individual properties that its members have, and those properties, as we shall see, are said to derive from the members' genes, which are, of course, individual. See Richard C. Lewontin, *Biology as Ideology: The Doctrine of DNA* (New York: HarperPerennial, 1992), 89, 93. If human beings are comprised from nature it is important to remember that nature is not merely an addendum to humanity's constitution, it *is* humanity's constitution. Humans are limited biological beings and that influences their self-understanding, their religion and their science. Yet, human beings are also beings endowed with freedom and are not completely genetically determined—even though they are genetically influenced (to be discussed in chapter four). See Willem B. Drees, *Religion, Science and Naturalism* (Cambridge: Cambridge University Press, 1996), 1.

[70]Elliott Sober and David Sloan Wilson, *Unto Others: The Evolution and Psychology of Unselfish Behavior* (Cambridge, MA: Harvard University Press, 1998), 26. I found this idea through Thomas Jay Oord's book, *Defining Love*.

[71]Oord, *Defining Love*, 108.

[72]For a recent study of this, see Cronk and Leech, *Meeting at Grand Central*.

[73]Darwin, *The Origin of Species*, 240.

looked at the groups *themselves* and not the individuals *within* the group. Accordingly, for Darwin, this in some ways brought more problems than it potentially solved, which is why many sociobiologists are slow to agree that Darwin actually accepted group selection as a reasonable idea within evolution.[74] For instance, Michael Ruse claims that there was nothing "implicit about Darwin's commitment to individual selection. He had looked long and hard at group-selection and rejected it."[75] Ruse notes further that Darwin was an aggressive individual selectionist who unequivocally invoked individual selection. This penchant for individual selection is what pushes Darwinians to adopt a model for kin altruism in order to help explain why one might sacrifice for the group—in order to perpetuate what we might call "group DNA." However, the problem with the appeal to kin selection is that Darwin could not have possibly meant this, for the field of genetics was not combined with Darwinism until a century later, when the "modern synthesis" occurred.[76] Consequently, any reading of Darwin through a "genetic lens" would be anachronistic.[77]

---

[74]Cunningham, *Darwin's Pious Idea*, 30. Darwinism was originally thought to show that human nature was completely individualistic and selfish. Altruism, then, was regarded as the product of culture/environment alone, with "pure altruism" being unattainable. As I will show in this chapter, a better way to deal with altruism, however, is to provide awareness that the error of sociobiological fatalism does not rest in recognizing biological causality, but through minimizing the force of a multitude of other causal factors such as personal, cultural, economic etc. Here, human motivation in particular has multiple interacting causes that impact decision making, altruism or other subjective or moral actions—e.g., genes and memes. For more on this, see Pope, *Human Evolution and Christian Ethics*, 214. See also Pope, *The Evolution of Altruism and the Ordering of Love*, 105.

[75]Michael Ruse, "Charles Darwin and Group Selection," *Annals of Science* 37, no. 6 (1980): 620.

[76]Cunningham, *Darwin's Pious Idea*, 31. This was developed by Gregor Mendel; the "modern synthesis" is named after Thomas Huxley's grandson Julian Huxley's book *Evolution: The Modern Synthesis* (London: Allen and Unwin, 1942).

[77]In a way, the conversational milieu has not always made it easy to find a firm conclusion about altruistic behavior. In the "modern synthesis" or "neo-Darwinian synthesis," where genetics and natural selection are combined, small genetic changes are recombined and then ordered by the natural selection process. Again, it is important to note that genetics was not combined with Darwinism until a century later (and at that point dubbed "modern synthesis"). Gregor Mendel laid the foundations of genetics that ultimately helped R. A. Fisher, Sewall Wright and J. B. S. Haldane discover the "modern synthesis." Despite reading Darwin through a "genetic lens," advocates of the "modern synthesis" try to avoid anachronism by broadening the definition of what is meant by "individual selection," and often refuse to accept notions of group selection. Here, people subscribing to the "modern synthesis" accept that genes will pass on through individuals within the group. Thus, this stiff understanding of natural selection—relegated to individuals and isolated from group selection—limits the conversation concerning altruism, a

2.2 SOCIOBIOLOGICAL ALTRUISM:
FROM DARWIN TO DAWKINS

Through natural selection and environmental adaptability, replicator genes have been mutating and copying their way into the best-fit host from the beginning. If, then, humans and baboons have evolved by natural selection and a rigorous fitness survival game, then anything that has evolved by natural selection should therefore be selfish. Thus, we might mistakenly expect that when we look at the behavior of baboons, humans and all other living creatures, we will find that selfishness prevails in these creatures.[78] If this is the case, then one should erroneously expect all motivation, all behavior and certainly all phenotypic action to be entirely selfish, but this is not the case. Therefore, due to the complicated nature of the influences imposed on humanity, it is necessary to discuss, if only briefly, the genesis and journey of the evolutionary understanding of human beings.

It was not until the twenty-first century that scientists decoded the human genome. Yet, long before this momentous day, many asked how these genes shape human behavior and how humans function as a result.[79] One surprise in this decoding discovery was that there were fewer genes than once thought, providing more plasticity for cultural achievements.[80] This revelation made understanding the replicators more complicated due to the possibility of influences outside of biology. In the *Descent of Man*, Darwin notes,

> The moral nature of man has reached its present standard, partly through the advancement of his reasoning powers and consequently of a just public opinion, but especially from his sympathies having been rendered more tender and widely diffused through the effects of habit, example, instruction, and reflection.[81]

---

conversation that would do well to hear from both ideas of groups and individuals. See Cunningham, *Darwin's Pious Idea*, 40.

[78]Richard Dawkins, *The Selfish Gene*, new ed. (Oxford: Oxford University Press, 1989), 4.

[79]Robert Song, *Human Genetics: Fabricating the Future*, Ethics & Theology (London: Darton, Longman and Todd, 2002), 1.

[80]Holmes Rolston, "The Good Samaritan and His Genes," in Clayton and Schloss, *Evolution and Ethics*, 238.

[81]Philip Appleman, *Darwin*, 2nd ed., Norton Critical Edition (New York: Norton, 1979), 201.

Habit, example, instruction and reflection are all conditions of one's surroundings. These conditions tend to influence the host, possibly even
challenging the replicators' will.

Needless to say, there has been much quarreling about the extent to
which environment is significant for the phenotype—a debate that has
had a significant impact on the development of the theory.[82] A case in
point can be seen in Mark Ridley and Richard Dawkins's "The Natural
Selection of Altruism," in which they posit that "civilized human behavior has about as much connection with natural selection as does the
behavior of a circus bear on a unicycle."[83] What is more, E. O. Wilson
notes that human nature is merely a synthesis of genetic predispositions
that have been influenced by cultural norms and is a product of an evolutionary process.[84] Still another voice, Richard Alexander, author of *The
Biology of Moral Systems*, argues that natural selection only grants quasialtruistic acts that are actually disguised, self-interested forms of selfishness—nothing more.[85] In his article, "The Search for a General Theory
of Behavior," Alexander further states, "Society is based on lies. . . . 'Thou
shalt love thy neighbor as thyself.' But this admirable goal is clearly contrary to a tendency to behave in a reproductively selfish manner. 'Thou
shalt give the impression that thou lovest thy neighbor as thyself' might
be closer to the truth."[86] In the logic of *The Selfish Gene*, therefore, this
kind of impression is simply more productive and has a more fruitful
fitness for survival than in the case of altruism.

Seemingly good behavior and selfless action, then, is not as selfless as
it may seem. Wilson suggests that the function of even religious myths
and rituals is indoctrination to produce group loyalty, which, in turn,
produces group action and gives survival to all in the tribe.[87] People,

---

[82]I will critique these ideas in the subsequent chapter.

[83]J. Philippe Rushton and Richard M. Sorrentino, *Altruism and Helping Behavior: Social, Personality, and Developmental Perspectives* (Hillsdale, NJ: L. Erlbaum Associates, 1981), 32.

[84]Edward O. Wilson, *Consilience: The Unity of Knowledge* (New York: Random House, 1998),
260-61.

[85]Alexander, *The Biology of Moral Systems*, 3.

[86]Richard D. Alexander, "The Search for a General Theory of Behavior," *Behavioral Science* 20,
no. 2 (1975): 96.

[87]Edward O. Wilson, *Sociobiology: The New Synthesis* (Cambridge, MA: Belknap Press of Harvard
University Press, 1975), 565.

then, act with self-motivated altruism in their own genetic self-interest because it bonds them to others in their tribe while ensuring higher genetic survival fitness. Such self-preservation is the impetus for religious belief, subjugation and ritualistic group bonding. The more one is stable and connected, the higher the genetic fitness.[88] Here, humans may be genetically inclined to behaviors that lead to a kind of xenophobia driven by selfish motivation.[89]

One has to acknowledge, however, that the environmental influences that may impact a human organism potentially do so in the midst of a stasis period of biological evolution.[90] For example, David and Marvalee Wake and Gerhard Roth conducted a study on salamanders in which the authors combat the notion that salamanders only ate moving prey. In the study, these salamanders were quarantined and (more or less) forced to eat stationary prey—and they adapted.[91] Their extremely fast adaptation demonstrates that whether in a period of stasis or punctuated equilibrium both factors of biology and environment cannot be diminished. Thus, as Gould's writings on anagenesis and punctuated equilibrium also prove, one cannot rule out sociobiological and environmental explanations for many aspects of animal behavior.[92]

In regard to the plurality of influences on altruistic behavior, those within the field of sociobiology have attempted to provide a theoretical framework within which behavior that appears to result partly from nonbiological factors can be understood in evolutionary terms.[93] H. C. Plotkin puts it well:

---

[88]This idea of stability is found in Dawkins, *The Selfish Gene*, 13. He rightly suggests that the more stable the early organisms were the higher the chances of their survival. Because the early forms of life existed in a chaotic environment, stability invariably bred reproductive fitness.

[89]Wilson, *Sociobiology*, 249.

[90]Gould, *Punctuated Equilibrium*, 177.

[91]David B. Wake, Gerhard Roth and Marvalee H. Wake, "On the Problem of Stasis in Organismal Evolution," *Journal of Theoretical Biology* 101, no. 2 (1983): 211-24. See also David B. Wake and Gerhard Roth, *Complex Organismal Functions: Integration and Evolution in Vertebrates: Report of the Dahlem Workshop on Complex Organismal Functions—Integration and Evolution in Vertebrates, Berlin 1988, August 28–September 2*, Dahlem Workshop Reports (Chichester, UK: Wiley, 1989).

[92]George W. Barlow and James Silverberg, *Sociobiology, Beyond Nature/Nurture? Reports, Definitions, and Debate* (Boulder, CO: Westview Press for the American Association for the Advancement of Science, 1980), 257-69. See the chapter by Gould titled "Sociobiology and the Theory of Natural Selection."

[93]Pope, *The Evolution of Altruism and the Ordering of Love*, 102. In some ways, sociobiology is compatible with some version of natural law morality, though there are some areas of natural law that transcend pure sociobiological explanations. This is because it seems to provide much

Underlying all the biological and social sciences, the reasons for it all, is the "need" (how else to express it, perhaps "drive" would be better) for genes to perpetuate themselves. This is a metaphysical claim, and the reductionism that it entails is . . . best labeled as metaphysical reductionism. Because it is metaphysical it is neither right nor wrong nor empirically testable. It is simply a statement of belief that genes count above all else.[94]

This attempt at a solution is at the heart of Dawkins's endeavor to add to the biological discussion of selfish gene fitness with nonbiological arguments. Dawkins found it imperative to respond to the various arguments attacking the "biological approach" through the development of his "meme" theory.[95] By definition, memes are units of culturally developed ideas that impact behavior.[96] These are things such as soul, God, beauty and so on. According to Dawkins, similar to genes, memes also have a drive to survive.[97]

Dawkins argues that it is possible for memes to give humans the ability to resist the power of genes (e.g., voluntary celibacy with priests).[98] A gene for celibacy, by definition, cannot survive due to the fact that it cannot be genetically passed on to kin; the sociobiologist, however, must account for its continuation. Thus, it may be transferred via memes. Some religious groups place a great deal of significance on the value of celibacy, notably its eternal significance. Therefore, the meme survives from one generation to the next, not through genetic means, but by memetic ways.[99]

Similarly, Dorothy Nelkin is also convinced that the biological basis of human nature is influenced by the environment.[100] Human action is

---

evidence for drives for self-preservation that fit nicely within Thomistic natural law morality. See Boyd, "Thomistic Natural Law and the Limits of Evolutionary Psychology," 235.

[94]Henry C. Plotkin, Evolution in Mind: An Introduction to Evolutionary Psychology (Cambridge, MA: Harvard University Press, 1998), 94.

[95]It should be noted that meme theory is highly controversial even within the field of sociobiology and I do not intend to spend much time on it; yet, it is necessary to briefly acknowledge it.

[96]Dawkins, The Selfish Gene, 203.

[97]Ibid., 180.

[98]Ibid., 203-5.

[99]Boyd, "Thomistic Natural Law and the Limits of Evolutionary Psychology," 228.

[100]Dorothy Nelkin, "Less Selfish Than Sacred? Genes and the Religious Impulse in Evolutionary Psychology," in Alas, Poor Darwin: Arguments Against Evolutionary Psychology, ed. Hilary Rose and Steven P. R. Rose (New York: Harmony Books, 2000), 16.

often, possibly always, generated by mixed motives. This fact may be overlooked by those who want to reduce emotional and psychological complexity to a single egoistic motivation generated by selfish genes, as I will further discuss in chapter three.[101] Yet, admitting the prevalence of selfish predispositions does not mean that humans have to submit to a theory of universal "psychological egoism," the claim that people always pursue what they think is in their own self-interest.[102]

Just as nonbiological factors influence altruism, along with the complicated nature of how altruism works with the selfish gene, some sociobiological explanations qualify what the word *altruism* actually means. Ruse divides altruism into two terms: "weak agape" and "strong agape."[103] "Weak agape" is consistent with our evolved emotional predispositions and "strong agape" runs so contrary to our nature that it is unacceptable, irresponsible and even "morally perverse."[104] The "strong interpretation" means that one loves his neighbor only as one loves oneself (and only as an individual), whereas the "weak interpretation" expresses special preferences for self, family and friends. Here, Ruse holds ethics to be subject to what nature makes possible.[105] Similar to Ruse, Frans de Waal presents a view of altruism with significant limits. "Altruism," he states, "is bound by what we can afford. The circle of morality reaches out further and further only if the health and survival of the innermost circles are secure."[106] De Waal (and Ruse with him) would argue that it is better to trim the ethic down to a manageable size so that it can sustain altruism within the small circle to which it has been adapted.[107]

Still, the definition of altruism—especially in regard to human altruism—when defined as action motivated by a concern for the welfare

---

[101]Pope, *Human Evolution and Christian Ethics*, 224.

[102]Ibid.

[103]To his discredit, Ruse does not offer a clear or sound explanation of what makes a position "strong" or "weak" in the first place. Consequently, the trouble with adhering to an entirely biological influence from nature over nurture is that, as sociobiologists like Ruse or Wilson seem to be advocating, the argument that ethical appeals to widespread responsibility are impossible to fulfill and therefore futile. See ibid., 232, 242.

[104]Michael Ruse, "Evolutionary Theory and Christian Ethics: Are They in Harmony?," *Zygon* 29, no. 1 (1994): 17, 19.

[105]Pope, *Human Evolution and Christian Ethics*, 230.

[106]Frans de Waal, *Good Natured: The Origins of Right and Wrong in Humans and Other Animals* (Cambridge, MA: Harvard University Press, 1996), 213.

[107]Pope, *Human Evolution and Christian Ethics*, 231.

of at least one other person, touches upon the issue of intention and motivation.[108] Stephen Pope argues a four-step process of how an act of altruism takes place.[109] First, a person must have a certain kind of psychological motivation to act on behalf of another. Second, the person may understand (to whatever capacity he or she is able) and make a judgment on the worthiness of the motivation—factoring in false motives and instincts. Third, the person decides to act on the basis of this judgment, which becomes the directing motive of the altruistic act. Fourth, the person intends to do something to pursue a certain course of action to obtain the desired altruistic act. This process of intention and motivation is done either consciously or subconsciously, and a host organism acts on the urges of the gene that provide direction. In theory, the replicator gene has selected cooperation through altruism, which provides more stability and gene fitness. I will be addressing this subject more fully in chapter four.[110]

Motivation, then, is the source of energy that activates behavior in the organism.[111] It is typically influenced by numerous factors, such as the physiological, cognitive, emotional, cultural and social. Sociobiologists often suggest that all people share the same genetic influences that comprise the underlying causes of all motivations. Some of these dispositions seem to be chosen consciously by the agent and not strictly determined by biology.[112] Genetic basis seems to be only one of multiple factors that influence an individual's particular motivational structure. A psychologically healthy and mature person may act for reasonably different reasons than someone who is not—this is something that is rarely taken

---

[108]This was noted earlier in the introduction, but it seems worth repeating here to highlight the importance of motivation: intention is a critical factor for altruism—especially if defined by altruistic love. See Thomas Jay Oord, "The Love Racket: Defining Love and Agape for the Love-and-Science Research Program," in *The Altruism Reader: Selections from Writings on Love, Religion, and Science*, ed. Thomas Jay Oord (West Conshohoken, PA: Templeton Foundation Press, 2008). For a more comprehensive look into intention, and fuller definition of altruistic love, see Oord, *Defining Love*.

[109]This process is taken from Pope, *Human Evolution and Christian Ethics*, 222.

[110]Thomas Jay Oord says, "Environmental dynamics beyond the [agent's] own body, such as political, communal, and ecological relationships, also greatly shape intentional activity. The claim that agents feel a variety of influences and yet remain genuinely free suggests that intentional action occurs in a context and with constraints." *Defining Love*, 18.

[111]Sober and Wilson, *Unto Others*, 199-200.

[112]Pope, *Human Evolution and Christian Ethics*, 223.

into consideration when sociobiology is simplified.[113] Intention, then, refers to an agent's mental plan for action.[114] However, the caution is that acts need not be solely self-sacrificial nor involve extraordinary levels of self-denial on the part of an organism to count as genuinely altruistic.[115]

In regard to motivation and intention, Elliott Sober and David Sloan Wilson describe how a Sudanese tribe of people called the Nuer who were dominant over a century ago had a complicated and costly social ordering system that bonded the group together.[116] The group closeness and structure in turn made them more dominant in the region, especially in regard to military campaigns. This work retrieves aspects of evolutionary thought that have fallen from grace with the preeminence of individual selection and gene-level selection in neo-Darwinism.[117] Sober and Wilson claim that it is likely that humans have evolved much for the benefit of the group,[118] a position that consequently "rejects simple egoism, simple hedonism, and simple pure altruism."[119] The discussion, then, becomes more complicated, having to somehow find a middle way.

The authors also claim that "natural selection based on cultural variation has produced adaptations that have nothing to do with genes."[120] In the same vein, Holmes Rolston suggests about group selection that "tribes of Good Samaritans will out reproduce tribes of thieves."[121] Rolston later argues that theories of altruism that are based on the genetic transmission of behaviors, and particularly the constraints on altruism that come about in these contexts, no longer hold when altruistic values are culturally or religiously passed on.[122]

Consequently, sociobiologists can provide grounds for a view of human nature that is not solely egoistic but takes into account genuinely

---

[113]For that matter, two psychologically healthy people might act for reasonably different reasons. See ibid.

[114]Sober and Wilson, *Unto Others*, 223.

[115]C. Daniel Batson, *The Altruism Question: Toward a Social Psychological Answer* (Hillsdale, NJ: L. Erlbaum, Associates, 1991), 97.

[116]Sober and Wilson, *Unto Others*, 186-91.

[117]Pope, *Human Evolution and Christian Ethics*, 220.

[118]Sober and Wilson, *Unto Others*, 194.

[119]Pope, *Human Evolution and Christian Ethics*, 221.

[120]Sober and Wilson, *Unto Others*, 337.

[121]Rolston, "The Good Samaritan and His Genes," 244.

[122]Clayton, "Biology and Purpose," 319. See also Rolston, "The Good Samaritan and His Genes."

altruistic, as well as selfish, motivations.[123] The conversation does not
have to be limited to biological influences on behavior. Environment,
then, can play a critical role in the evolutionary process (as will be discussed in chapter four). The fact that the environment by itself, on
certain occasions, can provide the ingredients required by the process of
natural selection gives environmental influences the status that critics of
biological determinism have championed.[124]

2.3 MOVING FORWARD

In brief, the purpose of this chapter was to introduce the basic sociobiological theories that attempt to explain the phenomenon of altruism. In
the following chapter, I will critique these explanations of altruism and
show how a better understanding of the whole person is needed to avoid
the reductionistic tendencies within the field of sociobiology, which
present a diluted understanding of the human person.

---

[123]Pope, *The Evolution of Altruism and the Ordering of Love*, 110.
[124]Sober and Wilson, *Unto Others*, 337.

3

ALTRUISM AND THE EXPLANATORY
LIMITATIONS OF EVOLUTION

◆

Sociobiological explanations of altruistic human action, especially those based on genetic evolution alone, are not fully satisfying. This chapter argues that, although biological explanations do account for portions of our understanding of altruism, there are also serious environmental dimensions at play in our decision-making processes. Furthermore, any altruistic action performed by individuals has a cascading effect, creating new patterns or norms in behavior, and leading to habit formation in the agent. Our genetic inheritance does not determine our future actions, and there is much in terms of human action that we—including sociobiologists—do not understand.[1]

Therefore, this chapter has three major critiques of sociobiological explanations of altruism. First, I will address the fact that sociobiological explanations of altruism *alone* do not completely explain the phenomenon of human altruism. Rather, the role of culture and its obvious influence on learned human behavior point to the reality that we are not merely the products of our genes. Second, I will discuss the issue of how sociobiologists often invoke problematic language when explaining altruism. This type of rhetoric exposes numerous inconsistencies among prominent sociobiologists. Third, there is an inability of sociobiologists to explain altruistic behavior without resorting to reductionism. The

---

[1]Richard C. Lewontin, *Biology as Ideology: The Doctrine of DNA* (New York: HarperPerennial, 1992), 3.

main problem with such oversimplification is that sociobiologists do not see the whole human person. These three critiques compose a false opposition for sociobiologists between philosophy/theology[2] and sociobiology, and is the subject of the last major section in this chapter. This false opposition causes sociobiologists to moralize outside the bounds of science as well as commit the naturalistic fallacy.

Human beings have many genetic and environmental behavioral influences that should not be oversimplified.[3] However, the critical question, usually omitted by sociobiologists, is not *whether* genes or the environment influence humans (we know, of course, that they do), but it is *how* these influential constraints interact with each other. Is one more dominant than the other or do they situationally trade dominance? Asking whether we are a product of our genes or environment, born or bred—or perhaps a sort of combination of both—leads only to generalized solutions that depend on the preferential weight given to either of the influences.[4] We need to move beyond the prevalent reductive model if we are to acknowledge that the interaction is much more complex than the division between the two warrants. Accordingly, this analysis lays the groundwork for the following chapter, wherein I develop the idea that human capacities and outcomes of behavior, like altruism, are "brought into being not by one cause or even by two (genes and culture) coercing the consequence, but, rather, by elaborate networks of constraint."[5]

3.1 THE ENVIRONMENT AND ITS INFLUENCE ON HUMAN BEHAVIOR

The first major criticism of sociobiological explanations of altruism revolves around the role of the gene and the influence of the environment. As we saw in chapter two, it is common for sociobiologists to posit ex-

---

[2]This false opposition is often drawn between what is sociobiological and what is beyond the physical. For my purposes in this chapter, I will be using words like *philosophy* or *theology* to represent suprascientific notions that sociobiologists tend to balk at.

[3]According to sociobiology, genes ultimately determine human behavior since they restrain the human condition within narrow biological boundaries. See previous chapter. See also John Bowker, *Is God a Virus? Genes, Culture, and Religion*, The Gresham Lectures (London: SPCK, 1995), 5. There are also influences of culture (memes).

[4]Ibid., 110.

[5]The following chapter concerns free will and constraints. See ibid., 103.

planations for why altruism has not been eradicated from our societies (and our genes), and how altruistic action can increase evolutionary fitness.[6] However, these justifications of altruistic behavior are almost always genetic in nature. Yet, no human is divorced from her or his surroundings. Regardless of how isolated one feels, culture and community heavily influence human behavior. All enter life as social beings immersed in a family, a state, a productive structure, and they view nature through a "lens that has been molded by their social experience."[7] This reality is what prompted Richard Dawkins to advance his memetic theory.[8] As one might expect, my criticisms in this section revolve around, but are not limited to, this theory.[9]

To this end, when behavior is described as selfish by sociobiologists,

---

[6]Conor Cunningham, *Darwin's Pious Idea: Why the Ultra-Darwinists and Creationists Both Get It Wrong* (Grand Rapids: Eerdmans, 2010), 28. There was once a question about how "lethal" altruism actually was for organisms. In other words, sociobiologists questioned if altruism was bad for organisms, rendering them "unfit" for long-term survival.

[7]Lewontin, *Biology as Ideology*, 3. It might be noted that the necessity of cultural influence can be illustrated by the reality that, only next to the death penalty, solitary confinement is considered one of the harshest forms of punishment. See Frans de Waal et al., eds., *Primates and Philosophers: How Morality Evolved*, The University Center for Human Values Series (Princeton, NJ: Princeton University Press, 2006), 5.

[8]Humanity is in some ways at odds with itself, divided by conflicting inclinations of both nature and nurture, which, through intentional effort and discipline, are somehow kept in harmony. I will expand on this word *intentional* in the following chapter. See Stephen J. Pope, *Human Evolution and Christian Ethics*, New Studies in Christian Ethics (Cambridge: Cambridge University Press, 2007), 268. Human behavior, then, is assuredly also influenced by one's environment—for even biologists saw the need to develop the memetic theory. Sociobiologists often imply that people seldom, if at all, transcend the evolutionary forces that previously shaped their hominid ancestors' characteristic behavior. See Stephen J. Pope, *The Evolution of Altruism and the Ordering of Love*, Moral Traditions & Moral Arguments (Washington, DC: Georgetown University Press, 1994), 111-12. On purely biological grounds, human action always functions on a selfish/selfless spectrum. However, one can pass on a set of ethical beliefs, mores or practices to one's offspring, enabling them to take on extreme characteristics of either altruistic or selfish actions. For these reasons, a critical look at sociobiological explanations of altruism is in line. See Philip Clayton, "Biology and Purpose: Altruism, Morality, and Human Nature in Evolutionary Perspective," in *Evolution and Ethics: Human Morality in Biological and Religious Perspective*, ed. Philip Clayton and Jeffrey Schloss (Grand Rapids: Eerdmans, 2004), 319.

[9]The irony, of course, is that up to this point in his book, Dawkins has been describing how there are no influences beyond the gene. The "all powerful gene" is what dictates and determines all phenotypic activity. Then, at the end of his book, with a mere few pages about memes, Dawkins seems to unravel his main thesis. Here, he states that culture also influences human activity. The term *gene-culture coevolution* might help to explain the interconnected nature of the genes and culture—a term that can be a helpful concept as we look at how culture also has considerable influence on our phenotypic action. It is in this dual process that both nature in genetics and nurture through environment influence humans. See Bowker, *Is God a Virus?*

it is unclear whether they are talking about genotypical or phenotypical selfishness, or some kind of combination of these different kinds of self-ishness.[10] Conceptual confusion occurs, leading sociobiologists to adopt a debunking and critical approach to the problem of altruism, which essentially "takes the altruism out of altruism."[11] Yet, the environment is part of the evolutionary process. The problem of nature versus nurture has not yet lost momentum. It is impossible to pin down what percentage of human action is biological and what percentage is influenced by environment. Ruse and Wilson argue for heavy biological influences, viewing the human sense of morality as a biological adaptation much like hands and feet.[12] They commit to the position of housing altruism and altruistic action in strict biological frameworks. Ruse holds the idea that altruism is a constructed notion made up by our genes. He states in "The Evolution of Ethics," an article coauthored with Wilson, that altruism "is merely an adaptation put in place to further our reproductive ends. Hence the basis of ethics does not lie in God's will . . . or any other part of the framework of the Universe. In an important sense, ethics . . . is an illusion fobbed off on us by our genes to get us to cooperate."[13] In a similar fashion, he also states, "human beings function better if they are deceived by their genes into thinking that there is a disinterested objective morality binding upon them, which all should obey."[14] By this he claims that feelings of "right" and "wrong," which humans perceive to be outside of biology, are in fact brought about by biological processes. Thus, for scholars such as Ruse and Wilson, there is no debate about nature versus nurture.

---

[10]Stephen Pope says, "The terminology of phenotypical and genotypical altruism distracts and confuses more than it clarifies. It might make biological sense to say that a rabbit eaten by a hawk has been altruistic, or that a grazing antelope is engaged in egoistic activity, but it makes no moral sense to use that kind of language when speaking about selfish or unselfish acts of human beings." Pope, *Human Evolution and Christian Ethics*, 226.

[11]Robert L. Trivers, "The Evolution of Reciprocal Altruism," *The Quarterly Review of Biology* 46, no. 1 (1971): 35.

[12]Michael Ruse and Edward O. Wilson, "The Evolution of Ethics," *New Scientist* 108 (1985): 50. See also Michael Ruse, *Taking Darwin Seriously: A Naturalistic Approach to Philosophy* (New York: Blackwell, 1986), 222; as well as Michael Ruse, "Evolutionary Theory and Christian Ethics: Are They in Harmony?," *Zygon* 29, no. 1 (1994): 15.

[13]Ruse and Wilson, "The Evolution of Ethics," 50-52.

[14]Michael Ruse and Edward O. Wilson, "Moral Philosophy as Applied Science," *Philosophy* 61, no. 236 (1986).

If nature is shown to have multifaceted influence on human action, then Dawkins—apt to prioritize biological explanations—attempts to elucidate the genetic drive for reproductive fitness found in nature. He describes the reasons behind seemingly moral actions, negating any purely altruistic motives. Even with this new adjustment,[15] Dawkins struggles with certain complexities of human nature and is not able to find a comprehensive explanation of altruistic behavior in biology alone. It is here where Dawkins, often in company with Wilson, steps into a realm beyond biology to offer explanations that will alleviate the dissonance of what he calls the "problem of altruism." In other words, the basic moral awareness of the distinction between right and wrong became "etched into the neural circuitry of the human brain";[16] according to Dawkins, this either happens for genetic or memetic reasons. In fact, Dawkins's arguments seem to move from empirical to environmental without much hesitation.[17] Thus, by the very nature of the memetic theory, Dawkins appeals to the environmental because it offers the possibility that a force outside the gene has impact on the human.[18] Memes and genes, as currently described and defined by Dawkins, with all their absolute control, cannot coexist within a human in any logical way. One cannot have two wholly influential entities—and so, Dawkins's logic breaks down.[19]

Predating Dawkins, Thomas Aquinas saw the basis for much of human behavior in the natural instincts that humans shared with other animals: reproductive behavior, care for kin and so on. But it is the *moral* sanctioning and forbidding of certain activities that make humans fundamentally different and unique from the rest of the

---

[15]Dawkins confesses that this new idea might seem like an "extreme view." And, in fact, it was revolutionary on many levels, one of which being a very comprehensive yet not complete explanation behind our phenotypic behaviors. See Richard Dawkins, *The Selfish Gene* (New York: Oxford University Press, 1976), 12.

[16]Larry Arnhart, "Thomistic Natural Law as Darwinian Natural Right," *Social Philosophy and Policy* 18, no. 1 (2001): 28.

[17]Craig A. Boyd, "Thomistic Natural Law and the Limits of Evolutionary Psychology," in Clayton and Schloss, *Evolution and Ethics*, 228.

[18]Ibid., 232.

[19]While memes for Dawkins are evolutionary explanations for the environment, and do not compete for "total control" over human behavior, they do contend for influence.

animal kingdom.[20] If this is the case, and if "sanctioning" and "forbidding" are special to the human race and are, in a memetic way, what makes us human, then one must conclude that some kind of pure altruism is at the very least possible. In regard to altruism, Mary Midgley states, "It is not the slightest use suggesting genetic engineering as a short cut. . . . Some might be tempted to suggest cloning admirably charitable people in the hope of getting a new race without narrowed sympathies. But this is to forget the effect of individual life and choice in the shaping of virtues. . . . Engineering for supercharity . . . is a realm of pure fantasy."[21] Thus, Midgley is critiquing the idea that one can co-opt nature and divorce it from culture, distinguishing her ideas from those who hold that humans are merely biological: upbringing, cultural surroundings and circumstance influence the altruistic behavior in humans.

Memetic theory, then, seems to be attempting to label the unscientific as "science" in order to make it more widely accepted. By doing this, sociobiologists concoct explanations of altruism that attempt to explain all selfless action under one theory. Midgley argues that the word *scientific* carries with it the sense of "academic excellence" that other groups (social sciences and humanities) do not carry.[22] This practice forces social scientists and humanists to make their reasoning look like science, and is why the memetic theory satisfies people when it applies scientific principles to thought and culture.[23] We might then ask whether or not culture is the sort of thing that divides up into units. Wilson states that culture must be made of units (or atomizable) because this is the way we as humans naturally think.[24] Midgley responds to such by saying,

> Again, this argument reproduces, in a reverse direction, the same mistake which Aristotelian physics made when it extended explanation by purpose from the human sphere to the sphere of inanimate matter. Stones do not have

---

[20]Boyd, "Thomistic Natural Law and the Limits of Evolutionary Psychology," 228.

[21]Mary Midgley, *Evolution as a Religion: Strange Hopes and Stranger Fears*, University Paperbacks (London: Methuen, 1985), 62.

[22]See Mary Midgley, "Why Memes?," in *Alas, Poor Darwin: Arguments Against Evolutionary Psychology*, ed. Hilary Rose and Steven P. R. Rose (New York: Harmony Books, 2000).

[23]Ibid., 71.

[24]Ibid., 74.

purposes, but neither do cultures have particles. The example of physics cannot justify imposing its scheme on a quite different subject-matter.[25]

As we will see shortly in the third major point in this chapter, when individuals attempt to explain all principles with one theory, problems of oversimplification become inevitable.

In similar fashion, Stephen Jay Gould critiques the "ultra-Darwinists,"[26] John Maynard Smith, Richard Dawkins and Daniel Dennett, by noting that they share a conviction that natural selection regulates everything of any importance in evolution, and that adaptation emerges as a universal result and "ultimate test" of selection's ubiquity.[27] Concerning Dennett in particular, Gould mentions that Dennett's argument is based on metaphors that all share the common error of assuming that conventional natural selection, working in the adaptationist mode, accounts for all evolution by extensions—including environmental influences on human behavior—so that "the entire history of life becomes one grand solution to problems in design."[28]

This idea is challenging to sociobiology for two notable reasons. First, Darwin himself strongly opposed the "ultras" of his day—choosing not to radicalize himself by trying to explain everything from "onething." Second, modern evolutionary biology—with new nonselectionist and nonadaptationist data from population genetics, developmental biology and paleontology—make contemporary times an "especially unpropitious time for Darwinian fundamentalism."[29] Jerry Fodor and Massimo Piattelli-Palmarini, in *What Darwin Got Wrong*, state, "Perhaps there are as many distinct kinds of causal routes to the fixation of phenotypes as there are different kinds of natural histories of the creatures whose phenotypes they are."[30] In a sense, simplification in the explanation, though convenient as it is, is not always philosophically or even scientifically precise.

---

[25]Mary Midgley, *The Myths We Live By*, Routledge Classics (New York: Routledge, 2011), 92. See also Midgley, "Why Memes?," 75.

[26]It is hard to determine here whether or not Gould is using the term as a pejorative. Yet this is his language.

[27]Stephen Jay Gould, "More Things in Heaven and Earth," in Rose and Rose, *Alas, Poor Darwin*, 86.

[28]Ibid., 91.

[29]Ibid., 86.

[30]Jerry A. Fodor and Massimo Piattelli-Palmarini, *What Darwin Got Wrong* (New York: Farrar, Straus and Giroux, 2010), 153.

In like manner, Rose states, "The problem with evolutionary psychology is that, like its predecessor, sociobiology, it offers a false unification, pursued with ideological zeal."[31] The central inadequacies of biological theorizing on which evolutionary psychology is based are simple reductions.[32] First, naked replicators are empty abstractions; DNA as a molecule cannot replicate by itself—it requires the appropriate protected environment of the cell. Second, the relationship between genes and phenotypes is not linear. Neither cells, nor organisms nor behaviors come complete when copied from DNA. Third, individual genes are not the only level of selection; rather, selection operates at the level of gene, genome and organism.[33] Rose provides an example of how a faster antelope would be more fit to outrun a lion, but the genes of enhanced muscles would not work well without other adaptations in the antelope (e.g., increased blood flow to the muscles). What is more, natural selection is not the only mode of evolutionary change and not all phenotypic characters are adaptive.[34] So for Rose, this simplistic explanation of the way genes work sounds good in theory, but is problematic in reality.

In addition, some further breakdowns have occurred regarding the interpretation of replicator genes since the time of Darwin. For instance, while it is true that genes are the units of inheritance, they are not the units of evolution. As Gabriel Dover says, "Biological functions are a complex mix of adaptation, exaptation, molecular co-evolution and adoption arising from the properties of turbulent genomes in turbulent environments."[35] There are no "units" of evolution because all units are always changing. Genes are integral in the evolution of biological functions, yet evolution is not about natural selection of "selfish" genes

---

[31]Steven P. R. Rose, "Escaping Evolutionary Psychology," in Rose and Rose, *Alas, Poor Darwin*, 247.

[32]The below are based on ibid., 254-60.

[33]Ibid., 257.

[34]I should acknowledge that there are a number of factors to consider about how various traits get passed on or weeded out by natural selection. The most aggressive debates are between *adaptationism*, where every characteristic in an organism has been adapted and selected for a particular function, and *genetic drift*, where mutations in a genotype do not get passed on due to small population or lack of reproduction (and the genetic code gets lost). And so, in the context of the current discussion, some genetic changes can occur but end up being neutral with regard to adaptivity.

[35]Gabriel Dover, "Anti-Dawkins," in Rose and Rose, *Alas, Poor Darwin*, 64.

specifically.[36] Still, many sociobiologists may have had sufficient reason to be dissatisfied with the lack of appreciation concerning the mechanism of natural selection. However, they often mistakenly promote the idea that natural selection is "for the good of the gene."[37]

This misstep becomes part of two erroneous assumptions of the selfish gene model. First, survival of the "optimon"—or the unit that will benefit from natural selection—is an important aspect of evolution. The optimon, not the organism, is the only true, self-replicating entity. The second assumption is concerned with the nature of certain evolved biological structures. For adherents to the selfish gene model, all structures are adaptations and all are "improbable perfections" that could only have come about by natural selection. This view naturally leads to the opinion that each species is "an island of workability set in a vast sea of conceivable arrangements most of which would, if they ever came into existence, die."[38] This viewpoint also does not recognize that there are more influences on organisms than genes alone. "It is Dawkins's dangerous idea," says Dover, "not Darwin's dangerous idea, which is seriously misleading. Theorists from diverse disciplines seem, unfortunately, quite happy to accept that evidence for a genetic contribution to complex human behavioural or morphological traits inevitably means evolution of that trait by a natural selection of selfish genes."[39] Accordingly, a subtle shift from Darwin to Dawkins took place in the contemporary evolution discussion.

Another question arises about the nature of such memes. If memes actually are parallel to selfish genes, then they must indeed be fixed units (just as genes). But most of the concepts mentioned are far from being characterized as immutable.[40] Midgley states,

> If memes are indeed something parallel to genes . . . if they are hidden causes of culture rather than its units—what sort of entities are these causes supposed to be? . . . Information is facts about the world and we need to know where, in that world, these new and causally effective entities are to be found.

---

[36]Ibid., 48.
[37]Ibid.
[38]Ibid., 49.
[39]Ibid., 51.
[40]Midgley, "Why Memes?," 76.

Without that knowledge, the parallel between memes and genes surely
vanishes and the claim to scientific status with it. Meme-language is not really
an extension of physical science but, as so often happens, an analogy which
is welcomed, not for scientific merit but for moral reasons, as being a salutary
way of thinking.[41]

Dawkins seems to struggle with consistent language and explanation.
The "new genetics" and the subsequent "new evolution" revolve around
features of modularity, redundancy, combinatorial permutations, molecular drive and molecular coevolution that "collectively place the organism and its functional evolution well out of reach of the Dawkins's
selfish 'optimons.'"[42] Instead, Dawkins's arguments are based on too
simple of evolutionary concepts—what science currently knows now
about evolution is much more complex.

One can see how behavior is influenced by multiple sources beyond
the genetic, some of which are environmental. Therefore, drives and
urges cannot be understood without including the environments in
which they develop. Consequently, an organism's capabilities are bound
biologically to genetic origins, evolutionary history and cultural location.[43] The more knowledge that is gained in the field of sociobiology,
the more it is clear that the explanation of altruism cannot be contained
in one overarching theory. As noted, Michael Ruse defined altruism as
the term for organisms giving to others at cost to themselves,[44] and it is
explaining this "giving" entirely through one theory where considerable
breakdown occurs.

3.2 PROBLEMATIC LANGUAGE

Of the three major criticisms in this chapter regarding how sociobiologists explain the phenomenon of altruism, the second critique is
closely related to the issue discussed in the previous section concerning environmental influences on human behavior: in trying to
explain all behavior under one umbrella theory, sociobiologists often

---

[41]Ibid., 77-78.
[42]Dover, "Anti-Dawkins," 56.
[43]Clayton, "Biology and Purpose," 325.
[44]Michael Ruse, "Evolutionary Ethics Past and Present," in Clayton and Schloss, *Evolution and Ethics*, 42.

evoke, whether intentional or not, problematic language. The effect of this practice is the undermining of sociobiological explanations of altruism.

There is a strange yet pervasive phenomenon in sociobiology where genes seem to take on a "life of their own" in the language that is used.[45] Human traits and language, even personification, are implored to describe activities of genes (and to some extent memes, as well). Clark points out that "even if genes desired offspring, they would not be strictly selfish: only philoprogenitive."[46] Yet genes cannot "desire offspring" just as computer servers cannot desire the Internet. Genes do not think on their own, they merely react. There are no such "suprascientific genes" but rather just DNA inside a cell.[47] Consequently, when dealing with the language of kin altruism, we find the discussion laden with personified genes that act out of gene fitness, when in fact they are instead preprogrammed entities reacting in a prescribed manner. This is a fundamental logical flaw among many sociobiologists. Dawkins uses problematic language like this when he discusses altruism with ambiguous and even self-conflicting tones, particularly in *The Selfish Gene* wherein he expresses his intent to define altruism (and therefore, by contrast, selfishness) behaviorally, not subjectively—desiring to remain free of concern for the "psychology of motives."[48]

E. O. Wilson appeals to the same kind of language. His understanding that one might be endowed with a "biological imperative," and thus be programmed with altruism for the benefit of the agent, might be a form of "hedonistic altruism," like a mother who enjoys sacrificing for her child. This understanding of altruism was noted in chapter two, but has numerous troubled presuppositions, which impact the rhetoric used to describe altruism. J. W. Bowker echoes that idea when he states, "It is clear that we are not born with '*a sense*' of anything like 'good and evil.' We are born as a developing process, in which the structures of the brain (themselves still, at birth, with considerable development ahead of them)

---

[45]Even the title of *The Selfish Gene* shouts this not-so-subtle ideology.
[46]See Stephen R. L. Clark, *Biology and Christian Ethics* (Cambridge: Cambridge University Press, 2000), 63.
[47]Ibid.
[48]Dawkins, *The Selfish Gene*, 4.

prepare us for characteristic behaviours."[49] While such a self-defending assessment certainly could be labeled "hedonistic altruism," it is quite possible that it could even be considered a cultural phenomenon, especially considering Bowker's claim that "development"—a word that implies some sort of cultural process—needs to take place. Regardless of whether we are born into "hedonistic altruism," it seems that many sociobiologists waver between claiming that there are *only* genetic inclinations or genetic *and* cultural proclivities (not to mention genuine freedom of choice). One can see just such a vacillation in Wilson's *Sociobiology*, where he states, "Murder and cannibalism are commonplace among the vertebrates."[50] Not only is this assessment anthropomorphic because animals cannot murder,[51] but also it demonstrates a sloppy use of moral language and confuses the learned capacities of organisms with inherent capacities, thus perpetuating the desire of sociobiologists to move freely between the biological and ethical in alarming ways.

To this degree, there is, in a sense, a general habit of moralizing the nonmoral human. Genetically speaking, humans, like animals, are inclined to survive in the fittest and most efficient way possible. If this is the case, any way this happens is acceptable. When dealing with altruism (and morality in general), Dawkins has a problem at the foundation of his argument. If there simply is no mightier force in this universe than the gene (a frequent Dawkins hyperbole), morality is *relative* to the needs of the gene. This becomes problematic when moral language is employed to argue that there are no morals. Take for instance the classic argument in George Williams's *The Pony Fish's Glow*, where he writes about the "harem polygyny" of Hanuman langurs:

> Dominant males have exclusive sexual access to a group of adult females, as long as they keep the other males away. Sooner or later, a stronger male usurps the harem and the defeated one must join the ranks of celibate outcasts. The new male shows his love for his new wives by trying to kill their unweaned infants. For each successful killing, a mother stops lactating and goes

---

[49]Bowker, *Is God a Virus?*, 110.
[50]Edward O. Wilson, *Sociobiology: The New Synthesis* (Cambridge, MA: Belknap Press of Harvard University Press, 1975), 246.
[51]This was a concept first brought to me by Pope, *The Evolution of Altruism and the Ordering of Love*, 101.

into estrous. . . . Deprived of her nursing baby, a female soon starts ovulating. She accepts the advances of her baby's murderer, and he becomes the father of her next child. Do you still think God is good?[52]

By using words like *murder* instead of *kill* and then posing the provocative question, "Do you still think God is good?" Williams assumes that these actions are contrary to "goodness,"[53] but to acknowledge the possibility of goodness is to acknowledge the possibility of morality, something that scholars like Williams are loath to do, despite their flippant use of such language.

Dawkins also expresses ambiguity in his thinking toward altruism, stating, "Let us try to teach generosity and altruism because we are born selfish."[54] Elsewhere, Dawkins mentions that "if there is a human moral to be drawn, it is that we must teach our children altruism, for we cannot expect it to be part of their biological nature."[55] His concern for teaching our children altruism arose in his response to argumentation about living in an "awful world" where one is not naturally unselfish. Dawkins seems to admit that we should somehow try to be unselfish for the sake of society at large.[56] Yet, this line of reasoning does not follow along the logic of the selfish gene. If humans (and other animals) should care only about gene fitness, there should be no reason to care for morality and its influence on society. While Dawkins might admit that there are moments where it is genotypically advantageous for an agent to act altruistically in society, he muddies the waters when he introduces the idea of educating for altruism, which is contradictory to his erstwhile, wholly biological explanations of altruistic behavior.

Dawkins makes strong claims concerning the single-minded and unsympathetic unaltruistic gene, which make sense when looking at its basic foundation. However, the real problem comes in his inconsistencies in thought when addressing altruism. He seems to be trying to avoid anthropomorphism, but cannot escape a universe guided by "something."

---

[52]George C. Williams, *The Pony Fish's Glow: And Other Clues to Plan and Purpose in Nature*, Science Masters Series (New York: BasicBooks, 1997), 156-57.
[53]Pope, *Human Evolution and Christian Ethics*, 11.
[54]Dawkins, *The Selfish Gene*, 3.
[55]Ibid., 150.
[56]Ibid., 3.

For example, he uses words like *blind* and *pitiless*, terms that only apply to beings that actually are capable of sight and mercy.[57] Dawkins expresses disappointment and outrage at the nature of the universe, but he is inconsistent and illogical with a natural world in which nonhuman organisms have no freedom.[58] Notice his take in *River out of Eden*:

> The world would be neither evil nor good in intention. It would manifest no intentions of any kind. In a universe of physical forces and genetic replication, some people are going to get hurt, other people are going to get lucky, and you won't find any rhyme or reason in it, nor any justice. The universe that we observe has precisely the properties we should expect if there is, at bottom, no design, no purpose, not evil and not good, nothing but blind, pitiless indifference.[59]

Dawkins's terminology functions contradictorily, using words like *prostitution* to describe some actions of the selfish gene in nature as "bad."[60] This logic is problematic. Yet, he seems to set the stage for others to continue with this inconsistency. Anthony O'Hear makes a similar statement to Dawkins's in *Beyond Evolution*, when he says that in nature, one can see "the total prostitution of all animal life, including man and all his airs and graces to the blind purposiveness of these minute virus-like substances [called genes]."[61] Such erratic logic and problematic language, where prostitution is bad even though it is natural, should give us pause and make us question the adequacy of sociobiological explanations of altruism.

3.3 REDUCTIONISM AND ITS RELATIONSHIP TO THE EXPLANATION OF ALTRUISM

After discussing the first two critiques of sociobiological explanations of altruism, unacknowledged environmental influences and problematic

---

[57]Pope, *Human Evolution and Christian Ethics*, 12.

[58]Christians, on the other hand, have been aware of the problem of pain for some time. They seem to know that not everything that happens in the universe is just (thus the *problem* of pain). This idea is present in the correction of retribution theology found in the book of Job.

[59]Richard Dawkins, *River out of Eden: A Darwinian View of Life*, Science Masters Series (New York: Basic Books, 1995), 132-33.

[60]Pope, *Human Evolution and Christian Ethics*, 13.

[61]Anthony O'Hear, *Beyond Evolution: Human Nature and the Limits of Evolutionary Explanation* (Oxford: Clarendon Press, 1997), 152.

language, a third and most important criticism revolves around the tendency of sociobiologists to utilize reductionistic reasoning and not acknowledge the whole human person.

There are several kinds of reductionism that can take place when dealing with the explanation of altruism, and it is important to carefully define each of them in order to avoid their pitfalls.[62] First, "epistemological reduction" advances from unprovable assumptions that "all phenomena of life, society, and of mind are explicable by a unified set of physical laws."[63] This kind of reductionism leads to numerous suprascientific questions such as the following: Where do these physical laws come from? Are they constrained by anything? Are there exceptions? With epistemological reductionism, the traits found in higher levels of complexity are explained entirely in terms of what is discovered on lower levels of complexity.[64] Whereas epistemological reductionism deals with levels of explanation, the second form of reductionism, "ontological reductionism," concerns the kind of entities that ultimately exist. Ontological reductionism occurs where more complex, higher-level traits or entities are seen as nothing more than a particular way in which simpler traits or entities are organized. In this reduction, one posits that the integrity of the whole is determined completely by the traits of its constituent parts.[65] Here again, ontological reduction is essentially a suprascientific position concerning terms of being, which sociobiologists do not realize because they believe they do not have suprascientific or philosophical commitments.

Yet, there is a third form of "reductionism" that might be acceptable, owing less to the idea of someone oversimplifying explanations of

---

[62]These kinds of reductionism can be found in chapter three, "Varieties of Reductionism," of Pope's *Human Evolution and Christian Ethics*. I will refer to this section frequently.

[63]Ibid., 61.

[64]This diminution can be seen in instances where biological events are fully accounted for in terms of less complex chemical reactions, such as biochemistry in terms of chemistry, chemistry in terms of physics and so on. See ibid.

[65]Ibid., 69-70. Even if sociobiology has trouble explaining human or animal altruism through sociobiological behavioral theory alone, naturalistic explanations can begin to contribute to the conversation by trying to explain altruistic behavior through two questions: How can we explain behavior, and what is the relationship between the explanations and descriptions of behavior on one hand and normative or prescriptive behavior statements on the other? For example, men may be predisposed to run off with women for reasons of reproduction, etc., but are not actually unavoidably programmed to do so by genetics. This exception goes against a deterministic line of reasoning.

altruism and more to categorizing them. "Methodological reductionism" is where natural sciences can explain the workings of physical, chemical and biological processes without recourse to nonscientific or suprascientific ways of thinking.[66] In light of this limitation, Christian ethics can and should accept the results of methodological reductionism without developing epistemological or ontological reductionism.[67] A nonreductionist reading of evolution that recognizes its inherent directionality is consistent with Christian theology and is often understood as God operating through "secondary causes" made possible by the evolutionary process. The account of human nature as constituted by emergent complexity helps one understand pieces of key notions in Christian ethics, particularly love of neighbor and natural law.[68] Essentially, with methodological reductionism, a research strategy where one breaks down complex wholes into their component parts can be actualized.[69]

Still, unhelpful forms of reductionism do not have to be inevitable. It is realistic to think that one can take human evolution seriously while denying the idea that natural selection has fixed dispositions of behavior that unwaveringly lead us to maximize our inclusive fitness.[70] Midgley again points out that sociobiology casts a "false light because it is 'reductive' in the sense of ruling out other enquiries and imposing its own chosen model as the only norm."[71] She claims that sociobiologists use "illicit inflation" to make their reductive points.[72] Consequently, this kind of oversimplification and reductionism that many sociobiologists, such as Wilson, Dawkins and even Ruse take part in, is not helpful in the quest

---

[66]See ibid., 56-61.

[67]Ibid., 56.

[68]See ibid., 7. This also comes into light with Neil Messer's take on theological ethics. He says in *Selfish Gene and Christian Ethics* that "a Christian theological engagement with discussion of evolutionary ethics, therefore, can both clarify and enrich those discussions by reframing arguments about altruism in terms of the biblical command to love our neighbour." See *Selfish Genes and Christian Ethics: Theological and Ethical Reflections on Evolutionary Biology* (London: SCM, 2007), 248.

[69]Such as understanding the mechanics of the heart in terms of pumps and valves—this is the basis of all scientific inquiry. See Pope, *Human Evolution and Christian Ethics*, 57.

[70]Philip Kitcher, *Vaulting Ambition: Sociobiology and the Quest for Human Nature* (Cambridge, MA: MIT Press, 1985), 402.

[71]Midgley, *Evolution as a Religion*, 154.

[72]Ibid.

for an explanation for altruistic behavior. Reductionism is neither inevitable nor necessary.

To analyze, provide examples of and critique unhelpful sociobiological reductions, I will break down this subsection in the following ways: first, I will address the problem that many sociobiologists only view altruism through the prism of the selfish gene theory; second, I will show how this kind of reduction encourages sociobiologists to miss the myriad of influences on behavior, causing them to misunderstand the whole human person.

First, sociobiologists have a difficulty explaining the phenomenon of altruism because they are often looking solely through the lens of the selfish gene theory. In Bert Hölldobler and E. O. Wilson's *The Superorganism*, they state that altruism is a "calculated cost that can be beneficial when the cost to the agent is relatively low."[73] As Stephen Pope suggests elsewhere,

> Wilson is convinced that sociobiology can explain the deepest nature and function of morality. Rather than a supernatural code delivered from "on high," or a "spark" of the divine lodged in each person's conscience, moral codes have originated because they serve the fitness interests of their adherents. . . . Specific norms, for example regarding marriage, property, or truth-telling, are accepted because they yield fitness benefits for those who adhere to them, or at least for those who promote them in others.[74]

Wilson develops his idea further by stating that membership in dominance hierarchies, for example, "pays off in survival and reproductive success,"[75] and that compassion "conforms to the best interests of self, family, and allies of the moment."[76] In like manner, Richard Alexander says that "generosity and altruism are older than Dawkins implies, and far more complex. I hypothesize that they are as integral a part of human nature as being 'born selfish.'"[77] Elliott Sober and David Sloan Wilson are

---

[73]Bert Hölldobler and Edward O. Wilson, *The Superorganism: The Beauty, Elegance, and Strangeness of Insect Societies* (New York: W. W. Norton, 2009), 23.
[74]Pope, *Human Evolution and Christian Ethics*, 251.
[75]Edward O. Wilson, *Consilience: The Unity of Knowledge* (New York: Random House, 1998), 259.
[76]Edward O. Wilson, *On Human Nature*, 25th anniversary ed. (Cambridge, MA: Harvard University Press, 2004), 155.
[77]Richard D. Alexander, *The Biology of Moral Systems*, Foundations of Human Behavior (Hawthorne, NY: A. de Gruyter, 1987), 139.

a bit more cautious, as well as consistent, in their understanding that humans are both genetically selfish *and* unselfish:

> Group selection does provide a setting in which helping behavior directed at members of one's own group can evolve; however, it equally provides a context in which hurting individuals in other groups can be selectively advantageous. Group selection favors within-group niceness *and* between-group nastiness. Group selection does not abandon the idea of competition that forms the core of the theory of natural selection.[78]

The point of Sober and Wilson is not to paint some falsified or rosy picture of universal benevolence;[79] rather, it is to show the realities of altruism in both selfish and unselfish ways. By these brief examples, one can see how genuine altruism is often examined in the realm of the selfish gene theory. The point of contention with standard sociobiology, as stated above, is that there is no room left for an agent to be purely altruistic, even if the agent does obtain reciprocal benefits of some kind (though reciprocation might not be intended either consciously or unconsciously). Essentially, then, it is in humankind's self-interest to encourage an ethic of self-sacrifice, duty and honesty because we benefit from living in healthy communities where people act civilly.[80] The key here is this capacity evolved as a more "successful strategy" than one narrowly focused on only the welfare of the self.[81] Stephen Pope counters this idea with the notion of human motivational plurality, which includes the capacity to take another person's good (or the good of the community) as an end in itself.[82]

With plausible alternatives, such as Pope's suggestion, sociobiologists do not have to only understand altruism through the selfish gene theory, which advances the idea that only "hypocrites and idiots" are altruists

---

[78]Elliott Sober and David Sloan Wilson, *Unto Others: The Evolution and Psychology of Unselfish Behavior* (Cambridge, MA: Harvard University Press, 1998), 9. Emphasis theirs. This idea is concerning within-group and without-group behavior. They go on to say, "Rather, it provides an additional setting in which competition can occur. Not only do individuals compete with other individuals in the same group; in addition, groups compete with other groups."

[79]Ibid.

[80]Dawkins, *The Selfish Gene*, 214.

[81]Pope, *Human Evolution and Christian Ethics*, 266.

[82]Ibid.

and that all motivation is instead derived from self-interest.[83] Such a strong position is far too reductionistic. Critiquing Wilson and Dawkins, Mary Midgley claims that when comments like "we are born selfish" stem from sociobiologists, we ought to be suspicious that they are attempting to diffuse the argument through gross oversimplification.[84] Still, Wilson splits his view of altruism into his own colloquial definitions of "hard-core" and "soft-core" altruism. The former is generous (for a sociobiologist), and implies that there might in fact be room for genuine moral altruism—a category in which he includes Mother Teresa—not motivated by desire for personal reward or punishment. The latter presupposes egoistically that one might be "cheerfully subordinate" to one's "biological imperatives."[85] Yet it seems that when Wilson is looking at altruistic behavior even under this divided lens, he tends to stray back and forth between the possibilities that altruism is either genuine and selfless or functional and thus quasialtruistic and somehow subconsciously hedonistic.

Epistemological and ontological reduction can also be seen in sociobiological explanations of how kin selection is related to altruism. There is a deep-seated assumption within the sociobiological community that humans are selfish and even "nasty" creatures compared to the *zoon politikon*, or social animal, that Aristotle claimed them to be.[86] This notion remains commonplace despite the fact that much of the assumptions made are untenable when looking at the bulk of human evolutionary knowledge. We have descended from highly social animals (i.e., monkeys and apes) and have been in give-and-take relationships since humanity's formation. Life in groups is not optional, but rather obligatory.[87] Sociobiologists must begin to ask how kin relationships (and reciprocal forms of nonkin altruism) factor in to selfless behavior, and this raises the question of whether it even matters if one creates gene fitness

---

[83]Clark, *Biology and Christian Ethics*, 63-64.

[84]Wilson, *Sociobiology*, 3; Midgley, *Evolution as a Religion*, 144.

[85]For an excellent discussion about this see Pope, *The Evolution of Altruism and the Ordering of Love*, 111. Also see Michael T. Ghiselin, *The Economy of Nature and the Evolution of Sex* (Berkeley: University of California Press, 1974), 166.

[86]De Waal et al., *Primates and Philosophers*, 3.

[87]Ibid., 4.

or gets something in return for altruistic actions. To answer this query, one can see that there is a seemingly typical strategy of reductionism into which sociobiologists tend to slip.[88]

In like manner, sociobiologists often claim that where innate traits are transmitted without benefiting their owner, they must have benefited close kin on average in some way.[89] Stephen Clark says for neo-Darwinists, "It is axiomatic that any pretence of loving concern for strangers must be a lie, since our genes will not allow us to squander resources that might instead assist the carriers of those genes. It is not necessary that we *want* to spread our genes throughout the population, but all that we do want must be meant to serve that goal (or else be a by-product of some dangerous mutation)."[90] This need not be the case, as Midgley points out. She states that the hypocrisy involved in pretending to be altruistic would indeed be impossible in the world sociobiologists (such as Wilson and Dawkins) talk about.[91] Furthermore, such a view is too much like the idea of fostering better gene fitness, which seems to be a weak generalization.

Take for example a study examined in Frans de Waal's *The Age of Empathy* that connects human behavior to primate behavior, displaying the connection between innate capacities for altruism. In this study, a capuchin monkey reaches through an armhole to choose between two differently marked tokens, while another monkey, physically separated from the first, looks on. The tokens can be exchanged for food, but in different ways. One token feeds both monkeys and the other token feeds only the chooser. Capuchins typically prefer the more prosocial token.[92] So, is this an altruistic action, since it involves no direct reciprocation

---

[88]As an aside, an interesting form of simplifying evolution in general can be seen through differing cultural lenses of evolution. For instance, in a capitalist society, evolution is viewed from a survival of the fittest perspective. Yet in more socialist countries (Latin American ones in particular), they say that evolution is one of the best examples of cooperation. Everything has to work together in harmony in order to survive. Take, for example, the pounds of microbes living in our bodies. We would die without them and they are considered part of us. For more on this, see Anthony Campolo, *A Reasonable Faith: Responding to Secularism* (Waco, TX: Word Books, 1983).

[89]Midgley, *Evolution as a Religion*, 142.

[90]Clark, *Biology and Christian Ethics*, 64. Emphasis his.

[91]Midgley, *Evolution as a Religion*, 148.

[92]Frans de Waal, *The Age of Empathy: Nature's Lessons for a Kinder Society* (New York: Harmony Books, 2009), 194.

and offers no benefit to the altruistic agent? If so, is that altruistic action in contradiction to the idea of "being born selfish"?[93] Here, sociobiology is functioning merely through evolutionary biological explanations applied to the level of not only physical traits but also behavioral traits showing that there *is* such a thing as Darwinian altruism (dying for the hive, etc.). Through this kind of altruism that benefits the agent, sociobiologists might try to explain the concepts of egoism and altruism on the biological level, but are unable to do so entirely or completely due to the existence of either nonkin or nonreciprocating altruism. While nonreciprocal forms of altruism, as discussed in chapter two, account for some of these factors as well, explaining altruism through the selfish gene theory alone seems wanting.

There are additional natural phenomena that pose a challenge to naturalistic explanations of altruism. A study done at Taï National Park showed that chimpanzees took care of those in the group with whom they purposefully lived. When a leopard injured some group members, others licked injured chimpanzees's wounds to remove dirt and waved flies away from the infected area. They were also mindful of the injured members by slowing down the travel speed in order to keep them with the group.[94] This purposeful group behavior, even when it put the lives of the healthy in danger, made sense when looking at the group benefit. However, because group members functioned more efficiently and safely as a whole does not mean that there was not an element of selflessness and sacrifice to stay in the group. There was opportunity for the chimpanzees to cut their losses, especially when given the danger of caring for the wounded.

Still more criticisms come from new research in the subject area. For example, an observation from de Waal's research demonstrates how if someone gives two monkeys vastly different rewards for the same task, the one who gets the lesser reward at some point simply refuses to perform. In our own species, too, individuals may reject some income if they feel the distribution is unfair. Yet, logic would say that since any income should beat none at all, this means that both monkeys and people

---

[93]These considerations seem less obvious when applied to nonhuman animals.
[94]De Waal, *The Age of Empathy*, 7.

fail to follow, to the letter, the "profit principle."[95] By campaigning against unfairness, their behavior supports both the claim that incentives matter and that there is a "natural dislike for injustice."[96] It seems that oversimplifications, in regard to altruism especially, can backfire because the research goes both ways (as noted above in de Waal's example). This is precisely why sociobiologists need to be challenged in this area. In the words of Willem Drees, "We need a view of science which avoids understatement as well as overstatement."[97] Indeed we need a sociobiological perspective that explains altruistic behavior that does not resort to such straw man tactics.[98]

The second reason sociobiology alone is unable to explain altruism, without resorting to reduction, is due to its unwillingness to acknowledge the numerous constraints on behavior. This lack of recognition causes sociobiologists to misunderstand the whole human person. In essence, this problem exposes not just the downplaying of the importance of group settings for an organism, but also the *relationship* between biology and the environment. Darwinian biology shows how the human capacity for social order arises from social instincts and a moral sense shaped by natural selection in human evolutionary history.[99] Thus, this order is not

---

[95]One could also say "gene fitness" as well.

[96]De Waal, *The Age of Empathy*, 5.

[97]Willem B. Drees, *Religion, Science and Naturalism* (Cambridge: Cambridge University Press, 1996), 237.

[98]I agree with Neil Messer when he says in his *Selfish Gene and Christian Ethics* that it is "[not impossible] to give an account of genuine altruism within a reductionist frame of reference: one could say, for example, as [George] Williams does, that it is an accidental by-product of a boundlessly stupid evolutionary process, though the further questions . . . about why an accidental by-product of a boundlessly stupid process should have any claim on us also then arise." Here, Messer gives an acknowledgment of some sociobiological explanations while showing the inability of sociobiology to *fully* explain altruistic action. See Messer, *Selfish Genes and Christian Ethics*, 247.

[99]Larry Arnhart, *Darwinian Natural Right: The Biological Ethics of Human Nature*, SUNY Series in Philosophy and Biology (Albany: State University of New York Press, 1998). This debate is poignant in the case of human beings. Evolutionary thought has argued that biologically based capacities for empathy lead to a natural basis for extending human concern beyond one's immediate circle, ideas that are typically housed in the notions of "kin preference" and "reciprocity." A complementary side to that argument focuses on cognitive abilities for acknowledging the independent perspective of "the other" as fundamental to extending moral concerns, where Charles Darwin seemed to favor the extension of sympathy as a main motivating factor in developing a moral sense. He states in *The Descent of Man* that "the social instincts—the prime principle of man's moral constitution—with the aid of active intellectual powers and the effects of habit, naturally lead to the golden rule: 'As ye would that men should do to you, do ye to them

something that necessitates a biological impetus, and the tendency toward a false dichotomy between having to choose either environmental influence or evolution should be avoided.[100] Because some acts of cooperation and altruism seem to be biologically located in kin preference or some kind of reciprocity, one should not automatically assume that those explanations cover the totality of circumstances. What is more, merely because an altruistic agent can be the unintentional beneficiary of an act does not mean that she or he has to forfeit her or his altruistic status.[101] According to neurobiologist Steven Rose, sociobiologists consistently fail to distinguish these different uses of altruism and tend to lump together many different "reified interactions" as if they were all demonstrations of the "one character."[102] This seems to be an overly simplistic explanation. The contemporary conversation typically has an unclear definition of altruism and struggles to determine whether an act has to be completely other regarding, with no hint of self-concern, to be labeled as genuinely altruistic.[103] The ambiguity in this definition of altruism alone makes it difficult to factor in issues of kin preference and reciprocity.

Therefore, due to the interconnected nature of reciprocity and altruism, there is a difficulty in disentangling cases of motivational altruism that nevertheless enhance gene fitness. Altruism cannot be narrowly collapsed into a kind of shallow reciprocity ethic characterized by, as Robert Axelrod calls, "tit for tat" reciprocity.[104] This kind of reciprocity, incidentally, was largely the target of Jesus' criticisms.[105] In the New Testament, Christ discusses a kind of payback that is unexpected. Stephen Pope says this:

> Reciprocity is one form of prosocial behavior, it accounts for some but by no means all assistance giving not directed to kin; it comprises a significant but

likewise." Charles Darwin, *The Descent of Man and Selection in Relation to Sex*, 2nd ed. (New York: P. F. Collier, 1902), 194. See also Pope, *The Evolution of Altruism and the Ordering of Love*, 141.

[100]Pope, *Human Evolution and Christian Ethics*, 158.

[101]Ibid., 226.

[102]Steven P. R. Rose, *Lifelines: Biology Beyond Determinism* (Oxford: Oxford University Press, 1997), 281.

[103]Pope, *Human Evolution and Christian Ethics*, 227.

[104]Robert M. Axelrod, *The Evolution of Cooperation* (New York: Basic Books, 1984), 54.

[105]Pope, *The Evolution of Altruism and the Ordering of Love*, 120. See also Mt 5:38, 44, 46; Lk 6:35-36.

only partial subset of human social behavior. The egoistic presuppositions held by many sociobiologists makes them uneasy with the simple and straightforward claim that we have evolved emotional predispositions to help others, including both nonkin and nonreciprocators.[106]

It is, consequently, possible to assume that one cannot take on the concept of altruism without accounting for factors and influences outside of the biological. Richard Dawkins attempts to do such in "Memes: The New Replicators,"[107] where he seems to undermine his original argument of *The Selfish Gene* by suggesting that the gene might not be the only contributing factor to behavior.[108] This oversimplification is another error found in sociobiology that causes sociobiologists to see the human organism and not the human person and necessarily leads to a reductionist idea of how humans display altruistic action.[109] Even Dawkins would agree, especially in light of his idea of memetic influences, that humans are creatures beyond our genes.[110] To simplify and to diminish human behavior to only the self-perpetuating and uncontrollable urges of the gene, without any outside influence (such as the environment or free will) is problematic. We are capable of not only caring for our kin, but even also for enemies.[111]

One can also see that sociobiologists, especially those who are proponents of the selfish gene philosophy, have trouble explaining the acts of humans who often display extraordinary sacrificial, and consequently altruistic, behavior that reduces reproductive success without reciprocation or benefit to kin. This phenomenon can be seen through examples of a soldier falling on a grenade to save the life of her or his nonrelated comrades, celibate priests, purposefully nonprocreative couples sacrificing offspring for some greater good, single celibate humanitarian missionaries and so on. One might be able to argue that "falling on a grenade" could provide indirect reciprocation to the individual's group, and that,

---

[106]Ibid., 119.
[107]Dawkins, *The Selfish Gene*, 203-15.
[108]Ibid., 203. Dawkins focuses somewhat on "cultural transmission" and how it gives rise to a form of evolution.
[109]Pope, *The Evolution of Altruism and the Ordering of Love*, 120.
[110]Dawkins, *The Selfish Gene*, 204. Here, he describes various animals that have this memetic influence.
[111]Pope, *The Evolution of Altruism and the Ordering of Love*, 120.

in theory, the agent's phenotypical actions were genotypically selected through generations of benefit prior to her or his sacrifice. Yet, one cannot discount the idea that the same agent also has strong genetic urges to preserve her or his own life. Thus, one can see how the *whole person*, including free will, motivation and intention, need to be accounted for without reductionism.[112]

Sociobiologists like Wilson seem to vacillate on what accounts for altruistic behavior. For instance, Wilson talks about how "scientists and humanists should consider together the possibility that the time has come for ethics to be removed temporarily from the hands of philosophers and biologicized."[113] The foundation of Wilson's argument stems from the idea that the causal origin of morality justifies his reductionist point. In other words, Wilson uses a genetic fallacy to make ipso facto statements that diminish the influence of nurture.[114] As Stephen Pope observes,

> To attempt to account for all friendship and hatred, conciliation and fighting, peace and aggression, by reducing them to genetic interests and biological drives is to ignore characteristic features of human nature itself: the rich diversity of our social life, the communication of knowledge and moral wisdom through tradition, and the ability of human intelligence to creatively adapt to new conditions. The basis of such reductionism, in my judgment, is the sociobiological tendency to reduce all goods to one, inclusive fitness.[115]

Pope's explanation here is clear, and we ought to avoid the tendency to reduce all explanations of human altruistic behavior to the genetic without also taking very seriously a host of environmental influences. For example, there is often an inclination from parents of twins to make them as similar as possible. They might be given names beginning with the same letter or dressed identically, just as twin conventions might

---

[112]See Jeffrey P. Schloss, "Emerging Accounts of Altruism: 'Love Creation's Final Law'?," in *Altruism and Altruistic Love: Science, Philosophy, and Religion in Dialogue*, ed. Stephen G. Post et al. (New York: Oxford University Press, 2002), 212.

[113]Wilson, *Sociobiology*, 562.

[114]Pope, *The Evolution of Altruism and the Ordering of Love*, 100. Here, Pope makes a convincing case about the tendency toward reductionism found in the sciences. Sociobiologists take heavy attacks from Darwinian evolutionists because they leave room for the influences of nurture upon nature.

[115]Ibid., 102.

even give prizes for the most similar twins.[116] Yet, it would be impossible to develop an all-inclusive understanding of how the role of genes precisely relates to cultural influences on human behavioral variation.[117]

A further example of how oversimplifying altruism misses the whole human person lies in the assumption that altruists often have a dubious motive. It is certainly true that having companions and relations of connection offer immense advantages in finding food, protection from predators, and basic thriving and surviving tactics.[118] Nevertheless, this does not mean that all contact with anyone besides "self" is some kind of dually motivated interaction, yet the tenor for many in the sociobiological community would suggest otherwise. The quotation by Michael Ghiselin speaks for itself:

> No hint of genuine charity ameliorates our vision of society, once sentimentalism has been laid aside. What passes for co-operation turns out to be a mixture of opportunism and exploitation. The impulses that lead one animal to sacrifice himself for another turn out to have their ultimate rationale in gaining advantage over a third; and acts for the good of one in society turn out to be performed to the detriment of the rest. Where it is in his interest, every organism may reasonably be expected to aid his fellow. Where he has no alternatives, he submits to the yoke of communal servitude, yet given a full chance to act in his own interest, nothing but expediency will restrain him from brutalizing, from maiming, from murdering—his brother, his mate, his parent, or his child. Scratch an altruist, and watch a hypocrite bleed.[119]

Certainly this *can* be true, but is not rigidly true "all the time," for there are countless examples of those who selflessly give and do not receive enhanced gene fitness. What is more, sociobiology frequently ignores the union of affections that show the mutuality of friendship in contrast

---

[116]Leon J. Kamin, *The Science and Politics of I.Q.* (Mahwah, NJ: Lawrence Erlbaum Associates, Inc., 1974).

[117]Lewontin, *Biology as Ideology*, 33. It is true that heritability explains certain traits when looking at the population or statistical level; and no one could possibly predict how millions of humans might interact with individual cultures. And so, no concrete understanding of this will ever likely be possible.

[118]C. P. van Schaik, "Why Are Diurnal Primates Living in Groups?," *Behaviour* 87, nos. 1–2 (1983); Richard W. Wrangham, "An Ecological Model of Female-Bonded Primate Groups," *Behaviour* 75, no. 3–4 (1980).

[119]Ghiselin, *The Economy of Nature and the Evolution of Sex*, 247.

to the mere reciprocity of two self-interested individuals.[120] This complicated problem can be seen in contemporary examples of those who
remain celibate for philosophical reasons (such as individuals who want
to raise adopted children rather than genetic progeny). Here, individuals
might express altruism, but a positive environment for gene fitness is
neither consciously nor subconsciously the end goal. In short, sociobiology alone has an inability to explain altruistic behavior because it
cannot look past the selfish gene model and does not take into account
the complicated makeup of the whole human person.

3.4 A REDUCTIONIST-DRIVEN FALSE OPPOSITION BETWEEN PHILOSOPHY/THEOLOGY AND SOCIOBIOLOGY

As we have seen, sociobiologists can be critiqued on three levels: first,
they fail to give an adequate account for how the environment influences
altruistic behavior; second, they frequently invoke problematic language
and inconsistent logic when trying to explain altruistic behavior; and
third, they turn to reductionism that provides unsatisfactory explanations. These three points of weakness expose a false opposition between
philosophy/theology and sociobiology that sociobiologists adopt. This
section, then, will focus on the repercussions of the first three critiques
and show how sociobiologists dip into suprascientific notions, making
philosophical assumptions in their repeated and unsullied comments
and asides.[121] For our purposes in this book, as I attempt to connect
Wesleyan ethics to sociobiology, it is important to point out this dichotomy. In some ways, sociobiology's struggle to completely explain
altruistic behavior is due to this constant and all too frequent dabbling
into what is suprascientific. For those within sociobiology who are
trained in the physical sciences, this venture beyond science is outside
of their expertise.

Midgley's *Evolution as a Religion* comes as an attack on the many
evolutionary biologists who seem to effortlessly glide into suprascientific

---

[120]Pope, *The Evolution of Altruism and the Ordering of Love*, 119.
[121]For a critique of Wilson's *Consilience*, see Wendell Berry, *Life Is a Miracle: An Essay Against Modern Superstition* (Washington, DC: Counterpoint, 2000). I alluded to this dichotomy earlier in the chapter when critiquing Dawkins's ideas of memes.

discussions with unwarranted authority. For example, the last chapter of *The Selfish Gene* is one that Midgley critiques. Drawing the link from what is suprascientific and philosophical to the religious, she raises the question as to whether or not many sociobiologists are in fact making a religion of their own field. In her words, Midgley's point of the book is to "make us more aware of the underlying myths"[122] that dominate the discussion in this field. She also discusses how the language of the selfish gene is pervasive in the academy and its fervor and enthusiasm pitch toward levels of a religion. She believes that evolution is the first creation story to attempt to not have symbolism attached to it.[123] This attempt, according to her, does not seem to have been convincing, because the authors repeatedly slip into suprascientific language. Our theoretical curiosity is not detached from the rest of our lives—let alone how we look at the sciences[124]—and therefore we are never divorced from presuppositions. It is illogical and irrational to think that many sociobiologists are wholly above this. In this way, Midgley brings us the other side of Drees's *Religion, Science and Naturalism*, which states that religion is a "natural phenomenon" and acts as corrective. Still, there are times when Drees even acknowledges the difficulty in separating the suprascientific from the physical. He gives a quick warning against both downplaying the role of science as well as making "romantic and metaphysical interpretations of science."[125] Yet, this is generally not the tenor of his book.

Midgley criticizes the idea that when organisms act in an altruistic manner they are knowingly and perpetually purposeful in their selflessness. She says that the idea of the personified gene "is essentially pure fantasy, not only unsupported by the empirical facts which are supposed to be its grounds, but actually contrary to them, such as they are."[126]

---

[122]See introduction of Midgley, *Evolution as a Religion*.

[123]Ibid., 1.

[124]Ibid.

[125]Drees, *Religion, Science and Naturalism*, 2.

[126]Midgley, *Evolution as a Religion*, 3. See her attack pointed at Ghiselin, *The Economy of Nature and the Evolution of Sex*, 247. For further example, Matt Ridley is also guilty of slipping into the language of personified genes. Take the metaphor he uses in *The Red Queen*: "The thirty thousand pairs of genes that make and run the average human body find themselves in much of the same position as seventy-five thousand human beings inhabiting a small town." See *The Red*

Midgley also takes aim at whom she calls the "high priest"[127] of sociobiology, Richard Dawkins, arguing that his theory of the selfish gene, while biologically adequate, strays when it moves away from the physical realm. She says, "Individual motivation is only an expression of some profounder, metaphysical motivation, which [Dawkins] attributes to genes."[128] Pope, who has done much work at the intersection of altruism and evolution, argues a similar point in reaction to Dawkins's seamless transition from the physical to the philosophical in his explanation of selflessness, pointing out how Dawkins effortlessly extends the description of selfishness from genes to humans. Thus, Pope exposes Dawkins's conscious or unconscious leaning toward a suprascientific reasoning behind gene activity.[129]

As a result, Dawkins ends up lessening the strength of his sociobiological explanations of altruistic behavior when he steps out of the biological realm. If all behavior is fitness enhancing, how is it that humans practice behaviors that are genuinely altruistic, and how do we judge among the variety of natural impulses?[130] Both Wilson and Dawkins often explain altruism by funneling all moral principles into biological explanations, positing that human nature is simply the result of millions of years of surviving replicators. They suggest that humans should sometimes resist genes by appealing to the power of moral enculturation. Both for Wilson and Dawkins, the human agent's ability to resist the power of the biological[131] is a problem yet to be solved.[132] As such, there is a main difficulty with this stream of logic:[133] Dawkins says that "we no longer

---

Queen: Sex and the Evolution of Human Nature (New York: Macmillan Publishing Company, 1994), 92. Thus, his tendency is to find language that makes the gene out to be something a bit more complicated than it is in reality. Genes most assuredly have influence over their host, and drives rooted in genetic urges are a clear example of that. Yet when genes are personified into cognitive and thinking beings, the weight of the argument lessens. Yes, genes influence sexuality, yet sexuality is not *controlled* by some cognitively superior being called "the Gene." And if this is the case for sexual drives, might it also be true of nonsexual moral issues like altruism?

[127]This is in reference to a profile done of Midgley in *The Guardian*.

[128]Mary Midgley, "Gene-Juggling," *Philosophy* 54, no. 210 (1979): 455.

[129]Pope, *The Evolution of Altruism and the Ordering of Love*, 101.

[130]Excluding kin altruism, reciprocity etc.

[131]Biological issues like hearing failure or biological urges like eating, sex etc.

[132]Boyd, "Thomistic Natural Law and the Limits of Evolutionary Psychology," 232.

[133]As an aside, the notion that human action is materialistically or genetically programmed ignores the fact that a great deal of suffering is not caused by either bad luck or nature but rather

have to resort to superstition when faced with the deep problems: Is there a meaning to life? What are we for? What is man?"[134] He wrongly presumes a radical but false dichotomy between religion and science as he equates religion with superstition. He mistakenly regards the idea of "God" as a cheap explanation given by theologians as an alternative to natural selection.[135] These presuppositions continuously creep into Dawkins's selfish gene argument and undermine any definitive sociobiological explanations to altruistic behavior, which ultimately leads altruism to be labeled a problem.[136] What is more, a solely naturalist view of science has its limitations and often presents humans with all their capacities as biological beings, yet with limited memories and limited ability for rational reflection.[137] If humans are limited in this way, science— as a "successful rational enterprise"—cannot be what it is; humans, therefore, are necessarily more than mere biological beings and have the capacity to reach beyond what can be naturalistically understood.[138]

In this vein of mixing the religious (suprascientific) with the scientific (physical), Frans de Waal critiques what he calls the "Veneer Theory" of morality—the idea that moral behavior is only on the surface of a bad core,[139] a theory that can be traced back to Thomas Huxley, who thought that nature was a wild and cruel place. According to the Veneer Theory, we must act unnaturally and against our nature if we are to be altruistic.[140] Huxley likened humanity to a gardener who had to keep the weeds of nature out—we cut the grass to develop a fine, green lawn precisely because we do not like nature.[141] De Waal also rails against the misuse of the term *selfishness*, which has been plundered and removed from its true context, a context that necessarily includes knowledge of what one

---

by human irresponsibility, selfishness, opportunism or greed. Essentially, this oversimplification can lead to the notion that there is neither rhyme nor reason for human suffering. See Pope, *Human Evolution and Christian Ethics*, 13.

[134]Dawkins, *The Selfish Gene*, 1.

[135]For more on this, see Pope, *Human Evolution and Christian Ethics*, 13.

[136]The irony here is that for theistic evolutionists, evolution takes place *after* creation. It is simply irrelevant for evolutionary theory whether or not God was the first cause.

[137]Drees, *Religion, Science and Naturalism*, 238.

[138]Ibid.

[139]The following is taken from de Waal et al., *Primates and Philosophers*, 6-7, 100.

[140]Incidentally, this is the position Dawkins assumes throughout *The Selfish Gene*.

[141]De Waal et al., *Primates and Philosophers*, 6-7.

is doing. When humans or animals carry out an act, it cannot be labeled selfish without such knowledge; any time it is, one could just as easily say "self-preserving." What type of nature, if any, could one have in the absence of self-preservation? As Sober and Wilson state, "Egoists and individualists are objective, they suggest, whereas proponents of altruism and group selection are trapped by a comforting illusion."[142] Here, they wish to point to the "dark side" of altruism, but they are often not inclined to notice the "light side" of selfishness. As Christine Korsgaard points out, "It is not even clear that the idea of self-interest is a well-formed concept when applied to an animal as richly social as a human being."[143]

Geneticist George Price also worked out an equation for this kind of theory, with what is now known as Hamilton's Rule, by making many mathematical innovations based on its application.[144] His application of Hamilton's Rule suggested that the many behaviors we describe as altruistic are in reality selfish; in fact, one could take it to mean that there is nothing at all noble about them.[145] Hamilton himself described natural selection in terms of genes that were responsible for the evil in the world: "I believed that the violent and sadistic ideas, which seemed to arise so easily in my psyche in certain moods, must be vestiges from a period that occurred subsequent to the separation of the chimpanzee line."[146] In this way our actions are not part of some clean, rational intelligence but are rather steeped in our origins. If this is the case, surely we should analyze our actions with this apparent revelation in mind.[147] Yet, the stunning

---

[142]Sober and Wilson, *Unto Others*, 8-9.

[143]De Waal et al., *Primates and Philosophers*, 100.

[144]Concerning this idea of what is suprascientific and what is scientific, it is worth noting that Price regrettably took his own life in London; it was W. D. Hamilton who identified his body. The tragedy of this story can be found in Cunningham, *Darwin's Pious Idea*, 189. Hamilton described the situation like this: "A mattress on the floor, one chair, a table, and several ammunitions boxes made the only furniture. Of all the books and furnishings that I remembered from his luxurious flat in Oxford circus there remained a heap of clothes, a two volume copy of Proust and his typewriter." See W. D. Hamilton and Mark Ridley, *Narrow Roads of Gene Land: Last Words* (Oxford: Oxford University Press, 2005), 174. Conor Cunningham questions why Price killed himself, reasoning at last that Price thought he had discovered the formula for original sin. See Cunningham, *Darwin's Pious Idea*, 189.

[145]So Price, in hopelessness, took his own life with a pair of scissors to his throat.

[146]Hamilton and Ridley, *Narrow Roads of Gene Land*, 191.

[147]Cunningham, *Darwin's Pious Idea*, 189.

reality is that many sociobiologists themselves do not live life in a kind of suicidal nihilism as Price did. This has to cause one to question whether there is another way, a *via media*, that fills in the gaps.

As I will show later in this book, the eighteenth-century theologian John Wesley discovered a unique formula that speaks to this exact question of original sin and biological proclivities. Wesley, through his own method of behavioral constraints mixed with his intuition of the human biological condition, distinctly displayed how these seemingly disparate ideas do not have to be so distant. In so doing, Wesleyan ethics can have a strong premise with which to engage common ideas about the sociobiological possibility of altruism and the Wesleyan possibility of holiness. Consequently, it is the central aim of the later part of this book to explore the commonality between the supposed sociobiological problem of altruism and Wesleyan ethics.

3.4.1 Moralizing out of the bounds of science

If sociobiologists want to affirm that there is no place for connecting what is suprascientific to what is scientific, then understanding altruism in any moral sense is necessarily problematic for them. Due to the fact that evolution does not provide an ethical basis for promoting or restricting responsibility, sociobiologists who engage in such moralizing do so outside the bounds of science, and without justification within their own discipline.[148] This becomes most highly evident when the dialogue inevitably turns from the world of biology into what is suprascientific. However, it is in this interface between science and philosophy/theology where both the challenge and intrigue of science lies, with sociobiologists finding it necessary to move into this ambiguous territory when dealing with the complex problem of explaining human altruism.

For too long religion has kept science at bay. Drees states, "Religion is too important to leave to conservatives who attempt to save faith by keeping science at bay with the help of formal arguments, by rejecting science, or by replacing it with a reconstruction of their own."[149] Ironically, now it seems that science is guilty of doing the same thing. Even

---

[148]Pope, *Human Evolution and Christian Ethics*, 241-42.
[149]Drees, *Religion, Science and Naturalism*, 4.

worse, many in the field of sociobiology openly verbalize their desire to sideline religion, yet still find themselves slipping into suprascientific rhetoric. This is troubling not only because, as Drees points out, many sociobiologists seem to have a simplistic view of Christian scholarship—assuming all Christian scholarship to be from a conservative persuasion—but also because it is highly unproductive in regard to dealing with the explanation of altruistic behavior.[150] Pope offers a healthier perspective on the interconnected nature of both evolutionary biology and Christian theology and ethics—a perspective that is necessary in order to take a look at how altruism is composed in human nature. He says,

> While critical of evolutionary ideology, Christian ethics needs to engage evolutionary knowledge because it can help us better to understand important aspects of human nature and some of the enduring constituents of human flourishing. Christian ethics, especially as developed in the natural-law tradition engaged here, gives moral significance to the central constituents of human nature, so it must take seriously the massive body of literature and significant discoveries about where we come from, who we are, and what we need and desire as human beings. Knowledge of human evolution is a necessary source of insight for any contemporary Christian ethics that takes human nature seriously.[151]

Thus, a proper discussion between human evolution and Christian ethics can be profitable for both parties because of the increasing complexities in both fields.[152]

As a short aside, it must be said that important Christian figures, such as Pope John Paul II, took evolution, properly understood, to be compatible with the Christian view of human beings as responsible moral

---

[150]Drees's criticisms are not wholly valid when the discussion shifts away from mainstream Protestant Christian culture and some conservative theologians.

[151]Pope, *Human Evolution and Christian Ethics*, 5.

[152]So as not to distract from the flow and order of the main purpose of this section, it is important to note here that Pope takes an interesting look at the question of whether or not evolution is purposeless. If it is, that would undermine certain forms of Christian faith (a S. J. Gould argument). But, Pope says, according to biologist Simon Conway Morris, and astronomer William Stoeger, there is reasonable support for thinking that evolutionary directionality is consistent with a religiously based belief in the divine purpose of evolution. For more on the subject, see ibid., 111.

agents. Pope John Paul II's major reservation, however, came about from
a moral concern, that evolution should not be interpreted in such a way
as to downgrade human dignity.[153] This has to do with the longstanding
belief that a link to common ancestry with animals would be a threat to
morality.[154] These concerns often cause Christians to be cautious when
dealing with evolutionary biology. That being said, the intersection of
human evolution and Christian ethics is the "interdisciplinary zone . . .
where we must live," at least according to Holmes Rolston.[155] It is here
where the explanation of altruistic behavior resides.

One also has to be careful not to try to explain the totality of human
altruistic behavior under either science or religion. If Dawkins's point is
that we are exceedingly selfish—one might be able to substitute the word
sinful for *selfish* here—and that we need to be taught to behave better,
why invoke Darwinism to help prove something already well-taken?[156]
Cunningham states, "This whole rhetoric about a dramatic revelation of
inbuilt selfishness (we theologians call it original sin) and the call to be
jolly nice—is it there only to make it look tenable that there is such a
thing as a pure Darwinian world (where there is not)? This is a case of
smoke and mirrors. The reason we can call it smoke and mirrors lies
right there in *The Origin of Species*: evolution."[157]

3.4.2 Avoiding the naturalistic fallacy

When sociobiologists adopt a false opposition between philosophy/the-
ology and sociobiology, they are often tempted to make moral arguments outside the bounds of science, as noted above. This kind of moralizing frequently leads sociobiologists to commit the naturalistic fallacy
and derive an ought from an is.

There has been a propensity to describe natural phenomena, such as
altruism, generosity and so forth, and then ipso facto claim that all

---

[153]Ibid., 170.

[154]René Descartes, *Discourse on Method: Meditations on First Philosophy*, trans. Donald A. Cress,
3rd ed. (Indianapolis: Hackett Pub. Co., 1993), 33.

[155]Holmes Rolston, *Science and Religion: A Critical Survey* (Philadelphia: Temple University Press,
1987), vi.

[156]Cunningham, *Darwin's Pious Idea*, 247.

[157]Ibid. Though here, Cunningham rightly admits that Darwin does not use the term *evolution* in
this context. See Charles Darwin, *The Origin of Species*, The Harvard Classics (New York: Collier,
1961).

actions hereafter are merely natural and neither good nor bad. One cannot deduce moral conclusions from nonmoral, or "natural," premises. To explain the reasons for human behavior, whether influenced by biology or environment, and then proclaim that humans *should* function selfishly/selflessly is to commit a naturalistic fallacy.[158] Simply describing where something came from does not necessarily give justification for prescribing how one should behave in any moral sense. A better naturalistic description of science has to include social, empirical and rational aspects,[159] but also acknowledge humanity's freedom.[160]

Therefore, humans need a view of science that avoids understatements as well as overstatements. In other words, if the view of science is too modest, it is not relevant; and if the view of science is too pretentious, science itself becomes something suprascientific—which would make naturalism self-contradictory, as it would then be unable to "accommodate its most important contributing source, the natural sciences."[161] Yet, not many theologians have examined the evolutionary data for relevance to "loving thy neighbor."[162] Christians seem to be leery of diving into this intersection of love and evolution. This kind of love might in fact be where the answer to the unselfish selfish gene and an explanation of human altruistic action resides. If purged of improper reductionism, the Christian ethic of love can build on contemporary accounts of what is naturally human.[163] After all, love for God, neighbor and self are not fundamentally in competition.[164] Love takes place in a grander understanding of cooperation and caring for the common good.

---

[158]As Drees puts it, "A philosophical view of science can be considered both as a description *of* and prescription *for* [ethics]." Emphasis mine. Drees uses the word *science* at the end; but connecting it to the naturalistic fallacy makes more sense in terms of ethics. Drees, *Religion, Science and Naturalism*, 240.

[159]Ibid.

[160]I will expand this idea of "freedom" in the following chapter. Morality can often times be traced to evolutionary roots and the genetic makeup of humans. Religion, and the traditions that it is surrounded by, remain influential and important in human existence, yet not altogether otherworldly. As Drees states, there truly is a sense of beauty and unexplainable majesty to the natural world. This is a hallmark of his book (ibid.). There simply does not have to be a notion that religion and science are in some kind of epic battle where the stakes are high on both sides.

[161]Ibid., 237.

[162]Pope, *The Evolution of Altruism and the Ordering of Love*, xii.

[163]Pope, *Human Evolution and Christian Ethics*, 215.

[164]Ibid., 239.

Christian ethics can profit by recognizing the functional value of morality without presuming that morality is only meaningful for its social and biological functions, by noting that evolution has shaped important levels of other emotional and cognitive constitutions as humans.[165] This recognition would mitigate both reductionist tendencies and naturalistic fallacies. At what point morality actually burgeoned from social life is impossible to determine, and there is no comprehensive convincing argument that completely "explains" the origin of morality. We can surmise that emerging social conventions are based on various kinds of reciprocity. These norms bring a tendency to monitor compliance, retaliate against cheats and reward people accepting the mores of the culture. But it is unclear when the fear of punishment or "getting caught" was transcended by a higher calling or obligation to "do good."[166]

G. E. Moore, the father of the phrase *naturalistic fallacy*, found in his *Principia Ethica*,[167] has the sometimes overextended objection to "naturalistic ethics" that has much to do with the "Is-Ought" idea.[168] The widespread and entrenched objection to connecting the *is* of human nature with the *ought* of ethics is a major philosophical obstacle that tends to prescribe human behavior from biological explanations. Yet a healthier understanding of this is-ought idea would help develop a theory of the ordering of love that incorporates natural human affections.[169] One must be cautious about assuming that all reciprocation, whether kin or stranger, is wrong or naturally selfish. Ideas of kin preference are not defaulted to run contrary to Christian thought. Interestingly, kin altruism theory can also be seen as a resource that might be able to facilitate the retrieval of some Thomistic insights about the natural ordering of love. Here, recognition of the natural and social conditions is essential for a "love ethic" consisting of mutuality and equality.[170] There are also

---

[165]Ibid., 259.

[166]Ibid., 259-60.

[167]G. E. Moore, *Principia Ethica*, Dover Philosophical Classics (Mineola, NY: Dover Publications, 2004).

[168]W. D. Hudson, *The Is-Ought Question: A Collection of Papers on the Central Problems in Moral Philosophy*, Controversies in Philosophy (London: Macmillan, 1969).

[169]Pope, *The Evolution of Altruism and the Ordering of Love*, 158.

[170]Post et al., *Altruism and Altruistic Love*, 333-34.

copious texts in the Christian Scriptures that point to reciprocation in altruism that does not negate the altruistic act.[171]

Philosopher Holmes Rolston points out that the logic of sociobiology *requires* one to assume that the famous Good Samaritan is genetically unable to act in the victim's interest.[172] The Good Samaritan has merely a seemingly real concern for the victim. Rolston's criticism comes from the knowledge that biological traits alone do not tell us anything about the *moral* character of a given act. Pope adds to the argument, noting, "To say that a parent who dies saving his own drowning child is both phenotypically altruistic and genotypically egoistic is as ethically trivial as it is emotionally sterile."[173] Here, it is the difference between acting on choice (reading a book, running an errand, having a coffee with someone) and inadvertently acting (scratching an itch or tripping over something). Actions can only be understood in the *full context* of specific intentions, motives and beliefs of the agent. Thus, the sociobiological project of attempting to "explain" human behavior in only behavioral terms hinders its analysis of genuine human altruism.[174]

Michael Ruse, on the other hand, demonstrates—by comparing humans and ants—why humans have developed morals according to evolutionary theory. Ants are hardwired to be a certain way and have their own set of virtues where "there is no need for learning . . . but it comes at a high cost."[175] An ant will behave instinctively even when the environment changes. This single-minded action might jeopardize the single ant or the colony itself. "Generally," he goes on to say, "this does not matter, because ants are produced cheaply—a queen can afford the loss of a few thousand. Humans are beings that require a great deal of care and only a few can be produced."[176] Humankind needs the ability to respond to change in order to ensure the highest level of care, and having a moral sense gives humans the ability to do that. Ruse goes on to use

---

[171]The entirety of Matthew 6 (among others) is this way: Mt 6:4: give alms *"so that . . . Father . . .* will reward you"; Mt 6:15: forgive to receive forgiveness; Mt 6:33: striving first for the kingdom brings rewards.

[172]Holmes Rolston, *Genes, Genesis, and God: Values and Their Origins in Natural and Human History* (Cambridge: Cambridge University Press, 1999), 252. This is part of his 1997–1998 Gifford Lectures.

[173]Pope, *Human Evolution and Christian Ethics*, 225.

[174]Ibid.

[175]Ruse, "Evolutionary Ethics Past and Present," 42.

[176]Ibid.

the analogy that humans are like sophisticated rockets that once launched can adjust to moving targets, whereas ants are like cheaper rockets that cannot change direction once they have been fired.[177] Yet Ruse falsely assumes that there are no external factors that affect human organisms when dealing with the phenomena of altruism. To assume that there are no outside influences, besides the biologically driven adaptation to be altruistic in certain environments for the preservation of one's own genes, does not take into account purposeful celibates, for instance.[178] Here, sociobiology tends to slip away from the facts and into "what should be." In a sense, there is a regular slide from an *is* to an *ought*, no matter how many road signs are put up condemning the logical slip. Nowhere is it more prevalent than with sociobiologists, who often study human behavior and then, like E. O. Wilson regularly does, tell us to get in step with it.[179] It would be like generalizing about what most art has been and then deducing what it should be.[180]

The virtue of charity found in Scripture, for one, both inspires a higher quality in the intensity of love and deepens the most profound kind of friendship. The virtue of charity in an altruistic sense restores, purifies and heals humanity's "natural" human capacity for love.[181] In a way, this is a reliance on grace given from God. Consequently, it might be taken to suggest that scientifically established knowledge of human nature is irrelevant to Christian ethics. Yet, on the contrary, this is far from being the case because grace acts on and *within* human nature.[182] This dance between the ordering of the world in nature and the ordering nature of love, a grace given by God, constitutes the foundation of altruism. In Thomas Aquinas's *Summa Theologica*, there is a profound section regarding the extension of charity to one's neighbor, which includes

---

[177]Ibid.

[178]Thus, according to Ruse, humans should be satisfied with "weak agape." Pope balks at this relegation by spinning the famous Dawkins argument to say that "a bear might be trained to ride a unicycle, but no one can teach a bear how to drive a car." See Pope, *Human Evolution and Christian Ethics*, 242.

[179]For a fascinating read critiquing Wilson on this circular logic, see Berry, *Life Is a Miracle*.

[180]Charles Jencks, "EP, Phone Home," in *Alas, Poor Darwin: Arguments Against Evolutionary Psychology*, ed. Hilary Rose and Steven P. R. Rose (New York: Harmony Books, 2000), 44-45.

[181]Pope, *Human Evolution and Christian Ethics*, 229. Here, "natural" means "created" or "as God originally intended."

[182]Ibid.

"sinners" and "enemies."[183] Aquinas is not professing that altruism necessitates reciprocity, nor is he stating that it cannot be beneficial to the agent. Rather, Aquinas claims that we simply extend charity to others because of the general order of love that has been established by God. Social psychologist Daniel Batson seems to ignore Aquinas's theological framework and foundation of his argument, when he makes the misleading statement that Aquinas assumes that humans are, "at the heart, exclusively self-interested."[184] Aquinas says quite the opposite by stating that humans belong to a different order established in the gracious and distinct love of God.

It is important to note here that the Christian life includes both "unilateralism"[185] and "mutualism."[186] One should see the "unilateral" or "nonreciprocal" dimension of agape, as described by Gene Outka in both *Agape: An Ethical Analysis* and "Universal Love and Impartiality."[187] Here, giving love from the agent outward as well as reciprocity within a bilateral community, Outka gets to the heart of the Thomistic understanding of charity. One does not have to be exclusively giving or refrain from any reciprocity to be found in the divine ordering of love. The minimal foundation of agape is an unwavering commitment to serve, a concept that truly comes to fruition in genuine friendship. Therefore, Christian love is "participatory and mutual within the *koinonia* or fellowship of the faithful, although this circle is by no means 'hermetically sealed.'"[188] It is precisely this fellowship that inspires and empowers outreach to the "other,"[189] which, in turn, may or may not be on the receiving end of reciprocity. If reciprocity does occur, this does not negate the

---

[183]Thomas Aquinas, *Summa Theologica* (1274), II-II, 25,1.

[184]C. Daniel Batson, *The Altruism Question: Toward a Social Psychological Answer* (Hillsdale, NJ: L. Erlbaum, Associates, 1991), 3.

[185]Unilateralism is the claim that the ethical core of agape flows in one direction from the agent to the recipient.

[186]Mutualism is the notion that agape consists in bilateral friendship or brotherly love.

[187]Gene H. Outka, *Agape: An Ethical Analysis*, Yale Publications in Religion 17 (New Haven, CT: Yale University Press, 1972), chaps. one and eight. See also Gene H. Outka, Edmund N. Santurri and William Werpehowski, *The Love Commandments: Essays in Christian Ethics and Moral Philosophy* (Washington, DC: Georgetown University Press, 1992), 1-103.

[188]Stephen Garrard Post, *A Theory of Agape: On the Meaning of Christian Love* (Lewisburg, PA: Bucknell University Press, 1990), 13.

[189]Pope, *Human Evolution and Christian Ethics*, 243.

koinonia that has developed, and quite possibly could even strengthen it. Thus, Christian love ideally moves toward "friendship" as seen in the exemplary expressions of marriage and the family as well as other intimate relationships (i.e., Jesus and John, David and Jonathan) challenging the narrowness of egocentrism and nepotism and even the restraints of reciprocity.[190] In this way, one might be able to avoid the naturalistic fallacy by embracing ideas from philosophy/theology and sociobiology instead of constructing a false opposition.

3.5 CONCLUSION

Sociobiological explanations of altruism are found wanting due to three main criticisms: they do not adequately acknowledge how one's environment impacts one's altruistic behavior; they often invoke problematic language and logic when discussing the phenomenon of altruism; and they implore reductionistic arguments that lose sight of the whole human person. These three poor explanations lead many sociobiologists to adopt a false opposition between what is suprascientific and what is sociobiological. However, many within the field of sociobiology still dabble into suprascientific rhetoric, which allows them to moralize outside the bounds of science and commit the naturalistic fallacy.

Multiple important characteristics of evolutionary accounts of altruism can be understood without simplifying human evolutionary makeup.[191] For one, there needs to be an acknowledgement that altruism and other forms of social behavior are influenced by a multitude of factors including temperament, emotional state of the human, concrete environment, the history of reciprocity between the giver and the recipient, social roles and so forth.[192] It is important to remember that there is no single "gene of altruism" that affects all human behavior.

---

[190]Ibid., 248.

[191]All four concepts are discussed in Pope, *The Evolution of Altruism and the Ordering of Love*, 121.

[192]It is true that evolution has shaped human nature at the motivation and instinctual levels so that, as a species, humans are not habitually drawn to actions that are self-destructive or reduce gene fitness. Additionally, the concept of altruism in sociobiology relies on the belief that other-regarding reasons for action and conscious motives reflect the deeper working of unconscious selfish desires. Yet, outside factors beyond the selfish gene influence the human organism leading to the necessity for additional angles of study to be included within the definition of altruism. For more on this, see ibid., 112.

Sociobiology need not support a relentless and opportunistic egoism, but a more complex understanding of human nature that accounts for its genuine (even if limited) capacity for altruism.[193] Furthermore, kin preference and reciprocity can be understood as evolved emotional predispositions that provide the foundation and basis for prosocial human motivations and action. Simply by the fact that humans struggle to comprehend acts of altruism does not ipso facto mean that altruism is counter to human nature.[194]

Looking at the insufficiency of sociobiological explanations of altruism, the discussion of E. O. Wilson and the responses to his work seem lacking. The history of human biology, where one might be genetically inclined to certain selfish (or even selfless) acts, does not provide justification for future human action. Just because we know how and what might influence human behavior, does not explain in what ways humans *should* behave. Biology cannot, by definition, prescribe moral behavior. Even if biology did entirely explain why humans behave altruistically, it gives us no guide at all as to how humans *should* behave. As Midgley states, "The extraordinary thing about sociobiology is that, officially and properly speaking, it arises from the recognition of this sociality in advanced creatures and is simply a set of theories to account for it. Although its rhetoric treats that sociality as a myth, its theoretical task is to admit it as a fact and to make evolutionary sense of it."[195] There seems to be a major gap in the fundamental argument from sociobiologists. This is a chasm where one has to dissect this naturalistic fallacy in order to get to the bottom of altruistic action. Humans share a huge percentage of genes with other high primates, yet humankind functions quite differently on behavioral levels.[196] The brain, for example, has extremely stable developmental processes that involve genes in the preparation for certain behaviors.[197] If, in fact, our genes

---

[193]Ibid., 121.

[194]Clayton, "Biology and Purpose," 334.

[195]Midgley, *Evolution as a Religion*, 141.

[196]Bowker, *Is God a Virus?*, 111. Some say about 98 percent, yet this is a gross oversimplification.

[197]On some level, there is also an "ethological fallacy" that takes place when one sees animal sexual behaviors, as an example, and tries to connect that directly to humans without acknowledging outside factors on human behavior. See John Bowker, "Origins, Functions, and Management of Aggression in Biocultural Evolution," *Zygon* 18 (1983): 369.

are closely connected to other animals, that does not explain why our behavior is so different.[198]

Consequently, genes and the environment form a portion (not totality) of the network that prepares humans for behavior, which take degrees of control over its own self. It is precisely this concoction of genes and environment that creates the freedom of human nature to transcend the biological and cultural.[199] E. O. Wilson and Richard Dawkins, on the other hand, often stand in contradiction to this idea (and are even sometimes inconsistent with each other). Wilson in particular says, "If the brain is a machine of ten billion nerve cells and the mind can somehow be explained as the summed activity of a finite number of chemical and electrical reactions, boundaries limit the human prospect—we are biological and our souls cannot fly free."[200] So Wilson would see all behavioral sciences subordinate to the sociobiological (or gene influence). Yet this statement goes against Dawkins's own claim that humans "alone" have the power to rebel (read "free will") against the tyranny of the selfish gene.[201] Wilson seems to be oversimplifying to duck out of acknowledging there could be a basis for free will and Dawkins seems to be flip-flopping what he thinks influences human behavior. It is with this stark example that this chapter concludes, having demonstrated that sociobiology does not hold the complete explanation for human altruism.

We must also ask, then, what kind of constraints are on human altruistic behavior without resorting to reductionist propositions. Human evolution has left us with a "heterogeneous, mixed, and conflicted set of human inclinations and the goods toward which they move."[202] With this foundation, the genetic makeup of humans seems not to completely push

---

[198]Bowker, *Is God a Virus?*, 106.

[199]It should be noted that environmental influences are not the only factor in making humans different than primates. Humans may share 98.4 percent of their genes with primates, but humans also share 50 percent of their genes with bananas (as the saying goes). The issue is not what genes we have but how they are ordered. And so, genetic differences cannot be underemphasized. Still, my point is to show how environmental factors do, in fact, *also* impact behavior and traits, but are not the sole difference makers. Ibid.

[200]Wilson, *On Human Nature*, 1.

[201]In the endnotes of this chapter, he lumps Wilson together with him in his claim. Dawkins, *The Selfish Gene*, 201.

[202]Pope, *Human Evolution and Christian Ethics*, 268.

human action into selfishness or complete altruism—instead, the human person lies somewhere in between.[203] A more helpful explanation of altruism instead of reductionistic accounts has to include the ability of an agent to utilize free will to overcome biological and environmental constraints on altruistic behavior—the subject of the next chapter.

---

[203] Again, this is a point that Dawkins concedes to. Dawkins, *The Selfish Gene*, 203-15. Furthermore, there is no reason why a given human act cannot be altruistic in the ordinary sense and, at the same time, constitute an example of "genotypical altruism," promoting the reproductive fitness of another organism at the agent's own reproductive expense. See Pope, *Human Evolution and Christian Ethics*, 225.

4

OVERCOMING GENETIC AND ENVIRONMENTAL CONSTRAINTS ON ALTRUISM

◆

In previous chapters, I not only have demonstrated how biology has influenced the human condition but also have conveyed the inability of sociobiologists to fully explain altruistic actions. In this chapter, I will address how genetic and environmental constraints on an individual's behavior impact human freedom and responsibility. In other words, one cannot merely ask an individual to "be moral" or "be altruistic." Instead, individuals reside in many locations on the spectrum of biological and environmental constraint.

Regarding these boundaries and human freedom, every human being has a particular set of genetic material that has been passed down through generations of selection through reproduction. Those genetic traits that could not withstand the highly competitive process of transmitting their characteristics to the next generation would not survive. Consequently, humans are naturally constrained to behave in certain ways as a result of their genetic makeup. Humans respond to their instincts that urge them to carry out certain tasks, such as finding a mate to reproduce and continue the cycle. These highly motivating compulsions seem to go against, or at least compromise, the idea of free will.[1] But the question that will be explored in this chapter is how much human

---

[1] Matt Ridley, *The Red Queen: Sex and the Evolution of Human Nature* (New York: Macmillan Publishing Company, 1994), 4.

biology has been constrained by eons of human evolution, and how much free will is available to humans to overcome genetic urges brought about by natural selection. If some free will genuinely exists, then free will was handed to humanity through evolution[2] as a means to keep humans able to adapt to life's unpredictability (including negotiating the complexities of living in the world with other sentient beings). Humans, then, retain the capacity to use free will. This freedom can go against urges to reproduce in a positive environment, which would promote gene fitness. When faced with a decision of obeying our genetic compulsions and following our free will, especially in the case of altruism, one can see that there is much to overcome.

This chapter addresses the idea that although altruistic behavior is significantly influenced by biology, humans do in fact have the ability to overcome influences on their behavior. To show this, I will first look at some specific ways in which human biology constrains altruism. Some of these factors include the mind/body connection associated with neuroscience and genetic determinism, as well as examples of "controlling genes." I will demonstrate that, while these proclivities seem to have total control on an individual, they are in fact influences and not deterministic mechanisms. Next, I will look at how individuals can overcome such behavioral constraints, such as engaging one's free will to resist one's genes and overcome environmental barriers. Last, I will build a case that

---

[2]There could be unease among some Christians about my word choices here. In my view, God and evolution are not antithetical. There is no reason that God could not have used the process of evolution to create free will, or even morality in his creation (possibly as a reflection of part of God's character). One might look at it as a sculptor creating a statue. The statue was created through and by instruments. Evolution, in our case, is the tool God used to shape our morality. Furthermore, one might look at the "Baldwin Effect" to justify the claim that free will was impacted by evolution (as well as impacting evolution itself). Oord describes the Baldwin Effect succinctly by using a bison/horse example. Somewhere in the evolutionary past of the bison/horse, a herd of animals were threatened. Some animals from the herd chose to fight and the ones who survived were stronger and bigger. Some animals from the herd chose to run. Here, the ones who survived the attack were faster, lighter and more agile. The former became the bison and the latter the horse and, consequently, the herd split out of the free will of some of the animals. It is worth noting, that for these animals, the free choice I am describing does not necessarily mean a *conscious* free choice. The Baldwin Effect is in contrast to the Lamarckian idea of a *particular* giraffe that continuously stretched its neck and passed its genes to its offspring. Lamarck's concept has gone out of favor among biologists. For more on this, see Thomas Jay Oord, *Defining Love: A Philosophical, Scientific, and Theological Engagement* (Grand Rapids: Brazos Press, 2010), 130-31.

by embracing a better understanding of freedom and responsibility individuals have the capacity to make altruistic decisions in a genuinely free way.

4.1 THE DETERMINED HUMAN PERSON?

Some sociobiologists (Richard Dawkins, E. O. Wilson, and Paul and Patricia Churchland, especially) have contended that the "problem of free will" has long been settled by biology.[3] To them, human behavior is the product of biology, just as height and weight, even though it is often unclear how this connection is truly made.[4] The correlation of human freedom and biological determinism has to be understood within the interwoven context in which a person's experience as a free moral agent and their biological limitations exist[5]—a context in which the whole person needs to be considered. As A. R. Peacocke says, "Free choice is only possible at all in a milieu in which natural processes follow 'lawlike' regularities so that 'free' choices have broadly predictable outcomes. Choice would be illusory if all were totally unpredictable."[6] For instance, if a person lifts weights they are limited by the maximum weight they can lift. But by exercising near the top of her or his constraints, the "freedom to lift" gets pushed higher and higher. "Free choice," then, is best described as the faculty to choose which constraints will determine action, rather than overlooking those constraints.[7] In other words, the boundaries of choice actually engender more freedom.[8]

---

[3]See Stephen R. L. Clark, *Biology and Christian Ethics* (Cambridge: Cambridge University Press, 2000), 117. Markus Mühling does a good job cautioning about the deadends of both dualism and reductionism, showing the complexities of human nature. See Mühling, *Resonances: Neurobiology, Evolution and Theology; Evolutionary Niche Construction, the Ecological Brain and Relational-Narrative Theology* (Göttingen: Vandenhoeck & Ruprecht, 2014), 51. Mühling focuses on neurobiology and neurophilosophy—a bit out of the scope of this book—nonetheless, he offers good insight into the complexities of the human condition.

[4]Indeed, Dennett discusses this issue with his ideas of human consciousness. It is clear that the brain impacts actions, but how does human consciousness actually work? This is unclear. See Clark, *Biology and Christian Ethics*, 117.

[5]Stephen J. Pope, *Human Evolution and Christian Ethics*, New Studies in Christian Ethics (Cambridge: Cambridge University Press, 2007), 176.

[6]A. R. Peacocke, *Theology for a Scientific Age: Being and Becoming—Natural, Divine, and Human*, Theology and the Sciences (Minneapolis: Fortress Press, 1993), 75.

[7]Pope, *Human Evolution and Christian Ethics*, 183.

[8]This concept is in line with a compatibilist viewpoint, which will be discussed shortly.

However, some sociobiologists outright disagree with the concept of free will and place all action in the realm of the biological. Francis Crick states, "Your joys and your sorrows, your memories and your ambitions, your sense of identity and free will, are in fact no more than the behaviour of a vast assembly of nerve cells and their associated molecules. . . . You're nothing but a pack of neurons."[9] Ultra-Darwinist interpretations of both genetic and environmental constraints tend to remove any notion of free will.[10] But one has to ask the question how free will could be genetic if it could run antithetical to genes. How can one have free will if there is no real self, that is, a self with its own irreducible integrity?

Yet Darwinism has not always been associated with such deterministic denials of altruism. It was not until neo-Darwinism that "free will" became regarded as the illusion sociobiologists now take it to be.[11]

---

[9]Francis Crick, *The Astonishing Hypothesis: The Scientific Search for the Soul* (New York: Maxwell Macmillan International, 1994), 3.

[10]Biology is often constraining and can limit actions and control behavior beyond altruistic proclivities. For example, there is a gene found in some fruit flies called the "segregation distorter" that tries to monopolize the genetic pool during reproduction. For the best illustration of this, see Brian Charlesworth and Daniel L. Hartl, "Population Dynamics of the Segregation Distorter Polymorphism of Drosophila Melanogaster," *Genetics* 89, no. 1 (1978). Because each fruit fly gives half of its chromosomes to combine with its mate, they only pass on half the genetic code out of the total possible. If it adds more than half the code, its partner would have to reduce some of its chromosomes to compensate. Fruit flies with the segregation distorter have chromosomes that purposely kill off their partner's chromosomes in order to be able to retain as much genetic material as possible. The fly, for instance, will produce half as much sperm, but all containing the deadly gene. And this ensures more gene transmission within the mating process. See also Ridley, *The Red Queen*, 97-98. Another interesting example of biological traits that constrain behavior is found in a study done in 1992 by R. W. Beeman et al. Beeman sheds facinating light on animal behavior that has been influenced by gene promulgation. Some female flour beetles have a gene called "Medea" that kills offspring that do not inherit their own gene. Thus, one can notice the powerful force of biology at work in the beetle's behavior: "A previously unknown class of dominant, maternal-effect lethal M factors was found to be widespread in natural populations of the flour beetle, Tribolium castaneum, collected on several continents. Such factors are integrated into the host chromosomes at variable locations and show the remarkable property of self-selection by maternal-effect lethality to all hatchlings that do not inherit a copy of the factor itself. Offspring are rescued by either paternally or maternally inherited copies. The M-bearing chromosome is thereby perpetuated at the expense of its non-M homolog. M factors that map to different regions of the genome do not rescue one another's maternal-effect lethality. Factors expressing these properties are predicted to spread in a population, even in the absence of any additional selective advantage. Similar factors also occur in the related species T. confusum." See R. W. Beeman, K. S. Friesen and R. E. Denell, "Maternal-Effect Selfish Genes in Flour Beetles," *Science* 256, no. 5053 (1992). Such influences have great sway over the action of an animal. Here, the biology does not merely impact the behavior after the animal is alive; it influences the animal before conception.

[11]Pope, *Human Evolution and Christian Ethics*, 159.

Sociobiologists, and neurophilosophers in particular, argue that free will, as it is usually understood, simply does not exist and there is no possible way that the evolutionary process can produce a being that is genuinely free to make moral or ethical choices.[12] However, not only is it unnecessary to derive this notion from Darwinian evolution, the idea of "free will as illusion" does not fit into the common human experience that implies *genuinely* free actions—including acts of self-sacrifice. Even prominent figures within the field, such as Richard Dawkins or philosopher Daniel Dennett, go back and forth on the issue.

When looking at individual selection and the impact that beneficial characteristics have on a reproducing species, it is necessary to discuss the connection between the biological and the mental.[13] The ideas of materialists, like Paul and Patricia Churchland, can be extremely helpful in understanding the connection between our biology and decisions to act in an altruistic fashion.[14] Dennett, a materialist in his own right, explains how conscious decisions are often connected to biology. He says in *Consciousness Explained* that any concept of our mind that is disconnected from brain states, or concepts that view the mind as "composed not of ordinary matter but of some other, special kind of stuff," is a dangerous dualism that is not supported by evidence.[15] Dualism of this kind in which our body is separated from our actions—whether it be moral

---

[12]Ibid.

[13]There are clear connections between action and biology. Oxytocin, for instance, is a hormone that is released during both sexual orgasm and some altruistic actions (from self-sacrifice to sharing a meal). These are biological responses to moral behaviors. For more examples of this, especially in relation to a mother and child, see Sarah J. Buckley, *Gentle Birth, Gentle Mothering: A Doctor's Guide to Natural Childbirth and Gentle Early Parenting Choices* (Berkeley, CA: Celestial Arts, 2009), 101-4.

[14]Dennett is also a materialist in his own right. This sometimes gives him a more determinist perspective on the human mind, although he is technically a compatibilist. As a materialist, he believes that humans are evolved organisms that are lacking a spiritual self or immortal soul. Instead, borrowing a phrase from Dawkins, Dennett believes that humans are the product of the "blind watchmaker" of evolution. As a result, Dennett leans toward a determinist viewpoint, but does not completely hold it; he does not believe, for example, that there is at any instance only one possible future. See Daniel C. Dennett, *Freedom Evolves* (New York: Viking, 2003), 194. See also Richard Dawkins, *The Blind Watchmaker* (New York: Norton, 1986). It is worth noting that many Christians also believe that there is no "soul" in a dualist sense of the word. Rather, the person should be looked at holistically. Much of the Christian church recognizes the human condition this way. I will discuss this in the following chapter.

[15]Daniel C. Dennett, *Consciousness Explained* (Boston: Little, Brown and Co., 1991), 33.

decisions or trivial ones—does not factor in the complex nature of making decisions that lead to actions. Thus, in the case of altruistic actions, there must also be a strong biological connection, rather than actions made purely by the will. So the question comes about, how do we look at these good actions that are willed? Is it the biology that "wills" them through some kind of genetic determinism? If so, are people responsible for their biologically determined actions? If the biology of a human cannot be separated from its will to do altrusitic acts, then those actions are neither good nor bad. Patricia Churchland strongly rejects the premise of these questions. She says,

> Intuitions . . . are products of the brain—they are not miraculous channels to the Truth. They are generated in some way by nervous systems; they are undoubtedly dependent on experience and cultural practices, however hidden from consciousness the causes may be. That we can introspect their source is just a fact about brain function—about what is and is not conscious. It implies nothing concerning the Metaphysical Truth about what those intuitions tell us.[16]

If the case can be made for intuitions, it could be extended to notions of free will as well. Churchland takes this strong stand and is correct in her understanding that there is nothing wrong with observing cause and effect patterns in brain activity. Yet she seems to quickly dismiss how (or, perhaps more importantly, why) those patterns are initially expressed, because for her the moral law is a myth. She describes morality as a natural phenomenon controlled and compelled by natural selection and anchored in the brain while at the same time being shaped by environment and culture.[17] So, moral action, or in our case altruistic action, is not a product of will, but a product of genes with minor environmental influences. Thus, the belief arises that modern science can account for every mental and physical phenomenon using the same scientific system by which photosynthesis, reproduction and other biological phenomena

---

[16]Patricia S. Churchland, *Braintrust: What Neuroscience Tells Us About Morality* (Princeton, NJ: Princeton University Press, 2011), 190.

[17]Ibid., 191. Again, I am not so convinced that just because morality comes about naturally, or through evolution, that it necessarily makes it antithetical to some kind of suprascientific deity.

are explained.[18] One has to question whether this generalized accounting is dangerously reductionistic.

Richard Lewontin and others summarize the determinist viewpoint in a succinct way, which might be a helpful clarification for this discussion:

> Human lives and actions are inevitable consequences of the biochemical properties of the cells that make up the individual; and these characteristics are in turn uniquely determined by the constituents of the genes possessed by each individual. Ultimately, all human behavior—hence all human society—is governed by a chain of determinants that runs from the gene to the individual to the sum of the behaviors of all individuals. The determinists would have it, then, that human nature is fixed by our genes.[19]

In this idea, the programmed genes govern the individual[20] and humans themselves are merely flesh-covered automatons. Neurological events begin before an individual is cognitively aware of making a choice, and those events are simply a product of biochemical interchanges; all that one does is simply another determined action of that biology.[21]

Evolutionary psychology also has a similar perspective on the human condition and human action. According to Robert Wright, evolutionary psychology shows "how elegantly the theory of natural selection, as understood today, reveals the contours of the human mind."[22] Wright later goes on to say, "Understanding the often unconscious nature of genetic control is the first step toward understanding that—in many realms, not just sex—we're all puppets, and our best hope for even partial liberation is to try to decipher the logic of the puppeteer."[23]

A good example of determined connection between biology and

---

[18]Dennett, *Consciousness Explained*, 33. Paul Churchland frequently mentions that someday we will understand the brain so well that we will stop using "folk" words to describe emotions. "Love," for instance, will be known as a biological process.

[19]Richard C. Lewontin, Leon J. Kamin and Steven P. R. Rose, *Not in Our Genes: Biology, Ideology, and Human Nature* (New York: Pantheon Books, 1984), 6. Found in Pope, *Human Evolution and Christian Ethics*, 161.

[20]It is important to remember that despite personified language, genes can only perform the function they are "programmed" to do. Genes cannot have a personified "will."

[21]Clark, *Biology and Christian Ethics*, 118.

[22]Robert Wright, *The Moral Animal: The New Science of Evolutionary Psychology* (New York: Pantheon Books, 1994), 11.

[23]Ibid., 37.

behavior is found in the book titled *The Mind's I*, edited by Douglas Hofstadter and Daniel Dennett, in which Dawkins wrote a chapter titled "Selfish Genes and Selfish Memes." Dawkins states that it does not matter whether the action of an individual is perceived as altruistic or not, all behavior of animals is powerfully under the control of genes (if even in an indirect way).[24] He also calls animals "survival machines"[25] that are dictated by the "policy-makers" that we call genes, whose every movements are executed by the brain and the nervous systems.[26] Dawkins says that brains became more developed with every passing generation. Somewhere down the evolutionary path, the brains of animals took over more and more "policy decisions." This would inevitably lead to a (not yet realized) state where genes would give these "survival machines" one dominant policy of doing whatever is necessary to keep the genes alive.[27] For this all to make sense, one has to remember that an animal is going to have a biologically determined end, namely to replicate itself in the best fit way. In other words, an animal will only be able to live within the confines of the matter in which it dwells. Such matter happens to be composed of genes that are necessarily looking out for "their own best interest."[28]

---

[24]See Richard Dawkins, "Selfish Genes and Selfish Memes," in *The Mind's I: Fantasies and Reflections on Self and Soul*, ed. Douglas R. Hofstadter and Daniel Clement Dennett (New York: Basic Books, 1981).

[25]Celia Deane-Drummond makes an interesting statement about this phrase. Noting the duplicitous terminology Dawkins sometimes uses, she says, "Statements such as, 'we are survival machines' are inevitably value-laden." See Celia Deane-Drummond, *Creation Through Wisdom: Theology and the New Biology* (Edinburgh: T&T Clark, 2000), 206.

[26]Dawkins, "Selfish Genes and Selfish Memes," 142.

[27]Ibid.

[28]As noted in the previous chapter, this kind of language often seems personified in an effort to show the connection between an animal's biological determinism—as if genes could think about the future and what might be the best path for the future of their kind. This kind of language also speaks of a kind of biological coercion, which I will discuss later in the chapter. Genes are selected through evolution to dwell in the best-adapted host, only to be further selected down to a better host in future generations. And when animals (humans obviously included) share genes with a partner, those genes can get transmitted to offspring who compose an environment that is often more fit. In this way, humans do not have the total capacity to resist such powerful sexual urges for adaptability. As Ridley points out, "From the point of view of an individual gene, then, sex is a way to spread laterally as well as vertically. If a gene were able to make its owner-vehicle have sex, therefore, it would have done something to its own advantage (more properly, it would be more likely to leave descendants if it could), even if it were to the disadvantage of the individual. Just as the rabies virus makes the dog want to bite anything, thus subverting the dog to its own purpose of spreading to another dog, so a gene might make its owner have sex

In order to move beyond the limitations placed on the conversation by Dawkins and others—from the *total constraint* of biology to a more moderate position of biological *influence* on an individual's altruistic actions, or even a perspective where a person is free to live *within* their biological constraints—it is important to remember a few key concepts concerning biology itself. Behavior such as altruism is likely not the result of a single gene, a point that Stephen Pope repeatedly brings up. This kind of language is typically invoked out of convenience or in an effort to communicate an idea. As Pope says, depicting a "gene for" altruism or a "gene for" selfishness merely "refers to a discernible genetic influence on variation in two populations."[29] The assumption by the determinists that materialism is the only logical answer to the question of how humans make decisions and perform actions should be challenged. The mind, for determinists, is composed of merely matter and is thus nothing but a "physical phenomenon."[30] Therefore, any choice is actually not a choice at all due to its biological constraints. But as we shall see, there is reason to believe that individuals are not totally slaves to their biology.[31] Humans instinctually know that we have some sense of freedom in choice—we are not solely bound or determined. Nor do humans feel like they are "held hostage" to their genes—though of course this feeling may be grossly deceptive.[32]

Despite being adamant about the relentless and all-encompassing influence of the gene, Dawkins has a famous line at the end of his chapter on memes in *The Selfish Gene*, which certainly brings the possibility of

---

just to get into another lineage." See Ridley, *The Red Queen*, 95. This characteristic makes sense if one looks at genes as having been pruned by natural selection to settle into the most useful way of reproducing.

[29]Pope, *Human Evolution and Christian Ethics*, 163.

[30]Dennett, *Consciousness Explained*, 33.

[31]As Francisco Ayala argues in "The Difference of Being Human," "To choose between alternatives is genuine rather than only apparent." See Francisco Ayala, "The Difference of Being Human," in *Biology, Ethics, and the Origins of Life*, ed. Holmes Rolston, The Jones and Bartlett Series in Philosophy (Boston: Jones and Bartlett, 1995), 122. See also Matt Ridley, *The Origins of Virtue: Human Instincts and the Evolution of Cooperation* (New York: Viking, 1997), 176.

[32]Some sociobiologists, such as E. O. Wilson, suggest that humans are in fact deceived. Yet this claim is often unsupported by and lacks critical evidence. Questions arise such as the following: Where does my conscience come from? What is the cause of these inclinations? What causes the biological urges to originate? Dennett is typically helpful here. He argues for a more open idea of freedom.

free will into the discussion. In it he states, "We, alone on earth, can rebel against the tyranny of the selfish replicators."[33] This is certainly an optimistic tone that has, for good reason, provoked much skepticism due to its inconsistency with the rest of the book.[34] In *Not in Our Genes*, Richard Lewontin, Leon Kamin and Steven Rose expose the idea of genetic determinism held by people like Dawkins and Wilson:

> Brains, for reductionists, are determinate biological objects whose properties produce the behaviors we observe and the states of thought or intention we infer from that behavior. . . . Such a position is, or ought to be, completely in accord with the principles of sociobiology offered by Wilson and Dawkins. However, to adopt it would involve them in the dilemma of first arguing the innateness of much human behavior that, being liberal men, they clearly find unattractive (spite, indoctrination, etc.) and then to become entangled in liberal ethical concerns about responsibility for criminal acts, if these, like all other acts, are biologically determined. To avoid this problem, Wilson and Dawkins invoke a free will that enables us to go against the dictates of our genes if we so wish. . . . This is essentially a return to unabashed Cartesianism, a dualistic *deus ex machina*.[35]

Yet, Dawkins defends himself by stating that there is a false dichotomy between having to choose either genetic determinism or free will. Dawkins and Wilson claim not to be genetic determinists, stating that genes "exert a statistical influence on human behaviour while at the same time believing that this influence can be modified, overridden or reversed by other influences."[36]

Trying to move away from dualism, Dawkins claims that humans can be separate and independent enough from genes to rebel against them. Essentially, "we do so in a small way every time we use contraception."[37] Lewontin and others would agree with this, except that it goes against the whole tenor of the selfish gene argument. It is fine for Dawkins to argue this point in the endnote of his updated book, but that does not take away from its inherent self-contradiction, which

---

[33]Richard Dawkins, *The Selfish Gene* (New York: Oxford University Press, 1976), 201.
[34]See ibid., 331.
[35]Lewontin, Kamin and Rose, *Not in Our Genes*, 283.
[36]Dawkins, *The Selfish Gene*, 331.
[37]Ibid., 332.

remains unaddressed and is still problematic for Dawkins. If someone like Dawkins ends up agreeing that there is something approaching what could be called "freedom" that will or could accommodate altruism (provided one rebels against the selfish gene), then they have overtly joined the ranks of "dreaded creationists"—to use a phrase by Dawkins—and probably not for the first time.[38] This is because any threshold like this that is crossed—which in turn enables altruism—cannot be disentangled from evolution. In actuality, it is precisely the *result* of evolution. If not, then one is most seriously advocating a form of *special* creation.[39]

Even if one does have free will, or at least is said to, "it is just another example of a true lie: a useful idea, but simply another meme."[40] Most people promoting the memetic theory assert that free will is merely a false story that forms part of our culturally transmitted memes.[41] So if free will is simply a lie, or as Wilson calls it, the "illusion of free will,"[42] then why would the genes allow such a destructive lie, especially if genes are personified like Dawkins and Wilson often assume? It makes very little sense to rationally believe that humans lack freedom of rational action[43]—the contradictions are myriad. Thus, for example, three important questions should be asked:[44] First, is culture made up of information units that are transmitted from generation to generation? If this is the case, free will could sabotage the continuity.[45] Second, if these units

---

[38]Conor Cunningham, *Darwin's Pious Idea: Why the Ultra-Darwinists and Creationists Both Get It Wrong* (Grand Rapids: Eerdmans, 2010), 175.

[39]For individuals like Dawkins, their analyses rest upon the dualism of vehicle and replicator. "Consequently," Conor Cunningham argues, "deviations from Dawkins's static, rigid interpretation of evolution are always going to appear as aberrations, as forms of deviance, indeed as unnatural." Ibid.

[40]Ibid., 207.

[41]Susan J. Blackmore, *The Meme Machine* (Oxford: Oxford University Press, 1999), 237.

[42]Edward O. Wilson, *Consilience: The Unity of Knowledge* (New York: Random House, 1998), 131.

[43]A. Corradini, S. Galvan and E. J. Lowe, *Analytic Philosophy Without Naturalism* (London: Routledge, 2006), 177. See also E. Jonathan Lowe, "Rational Selves and Freedom of Action," *Personal Agency* 1 (2008): 179-99.

[44]These come from Cunningham, *Darwin's Pious Idea*, 208.

[45]Jablonka et al. actually suggest that "epigenetic variations are generated at a higher rate than genetic ones, especially in changed environmental conditions, and several epigenetic variations may occur at the same time." See E. Jablonka, M. J. Lamb and A. Zeligowski, *Evolution in Four Dimensions: Genetic, Epigenetic, Behavioral, and Symbolic Variation in the History of Life*, rev. ed. (Cambridge, MA: MIT Press, 2014), 142.

actually exist, would they be qualified as replicators? Third, if the first questions are in fact answered yes, does it make sense to analyze the transmission of these replicators using Darwinian logic? Cunningham argues that the answer to all three of these questions is actually no, because to answer simplistically in the affirmative does not take a strong enough look into the reality of free will among humans.[46]

4.2 THE HUMAN PERSON AS INFLUENCED BUT NOT DETERMINED

According to E. O. Wilson, humans are "predisposed biologically to make certain choices."[47] This view of biological determinism, though quite common among many sociobiologists, is a bit flexible in its language. Wilson seems to be arguing not for some kind of "all-encompassing" biological determinism, but rather some kind of compatibilism, where a complete account of physical causation is compatible with an account of free will in relation to the same events (i.e., simultaneous compatibility). Dennett also holds the view of compatibilism where free will and determinism are compatible after all.[48] We must ask, of course, whether this compatibilism leaves the human truly free or merely biologically determined (masquerading as compatibilism). Humans are free to the extent that they are not biologically coerced and there is freedom of spontaneity, but there is no room for what might be called a "freedom of indifference."[49] Criticizing Wilson's ambivalence in the matter, Clark makes a nonhuman analogy, saying that

> E. O. Wilson's story of the bee that, in a way, feels free, even though its behaviour is precisely predictable by anyone with the necessary clues (about the position of the flower patch, the hive, and the bee's load of nectar), is no more than a cuter version of the Stoics' stone that rolls downhill in obedience, as it might imagine, to its own desire to fall. "Falling freely" is, emphatically and obviously, not doing something that it is in our power to halt by the mere

---

[46]Cunningham, *Darwin's Pious Idea*, 208.

[47]Wilson, *Consilience*, 250.

[48]Dennett, *Freedom Evolves*, 98. When trying to understand compatibilism, it is helpful to note that all actions are "caused" not "coerced." In other words, a compatibilist might argue that all events are caused; it just so happens that humans cause some of those events.

[49]Clark, *Biology and Christian Ethics*, 118.

exercise of our will. It does not immediately follow that there is no room at
all for any other sort of "freedom."[50]

Yet whether or not the bee is bound to go and gather nectar, the bee
might not be *specifically* bound to gather nectar from a *particular* flower.
This kind of freedom, albeit limited, is still situated within the concept
of freedom.

For example, Ting Zhang et al. address some examples of these genetic
predispositions in a 2005 study. These predispositions, albeit strong, do
not create inevitable human actions. Their report shows an increased
proclivity to the effects of addictive substances that lead the agent to have
higher risks of addictive behavior.[51] Although the genetic basis for alcoholism is still in its early stages, research has pointed toward a direct
link.[52] Other studies link multiple genes (like a sort of group synergy)
impacting both alcoholism and tobacco addiction and use.[53] As Kent
Dunnington notes, some studies show a variation of the alcohol dehy-
drogenase gene increases the possibility of alcoholism.[54] In this case,
one can even see this genetic influence of alcoholism influencing identical twins and fraternal twins differently. In one instance, twins carrying
the same DNA and family history of alcoholism were less able to me-
tabolize the alcohol than fraternal twins without the same problematic
genes.[55] Yet, as the tenor of Dunnington's book attests, there are outside
factors that mitigate even the strongest genetic predispositions. In this

---

[50]Ibid.

[51]Y. Zhang et al., "Allelic Expression Imbalance of Human Mu Opioid Receptor (Oprm1) Caused
by Variant A118g," *The Journal of Biological Chemistry* 280, no. 38 (2005): 32618-24. I am indebted to Kent Dunnington for pointing me to this helpful study. See Kent Dunnington, *Addiction and Virtue: Beyond the Models of Disease and Choice*, Strategic Initiatives in Evangelical
Theology (Downers Grove, IL: IVP Academic, 2011), 21.

[52]For more on this, see L. Buscemi and C. Turchi, "An Overview of the Genetic Susceptibility to
Alcoholism," *Medicine, Science, and the Law* 51 (2011). See also Marcus R. Munafò et al., "Lack
of Association of Oprm1 Genotype and Smoking Cessation," *Nicotine & Tobacco Research*, (September 2012). Still, Munafò et al. would admit genetics has a role in explaining smoking cessation, if even a modest role.

[53]See Rachel F. Tyndale, "Genetics of Alcohol and Tobacco Use in Humans," *Annals of Medicine*
35, no. 2 (2003).

[54]Dunnington, *Addiction and Virtue*, 21. See also John I. Nurnberger and Laura Jean Bierut,
"Seeking the Connections: Alcoholism and Our Genes," *Scientific American*, no. 296 (2007):
46-53.

[55]Donald W. Goodwin, *Alcoholism: The Facts*, 3rd ed. (Oxford: Oxford University Press, 2000),
89.

way, genes clearly influence moral behavior, but do not bind a person to definite actions.

There seems to be a common misconception in regard to what it is to have selfish genes. Most sociobiologists do not think that human genes are the puppeteers of human behavior or that all human actions are predestined by genetic inheritance.[56] It is also important to note that genes do not dwell in isolation. There is constant activity between the environment and genetics that create potential (or even probable) actions, but not necessarily determined actions.[57] Thus, what factors can be constituted as "genuine altruism"[58]—or even what that might look like—is not entirely set in stone. As Stephen Pope explains,

> Even on the micro-level, it makes no sense to assume a genetic determinism according to which genes by themselves somehow cause behavior. Genes never function as isolated causes of behavior but . . . rather as essential components of complex networks. Behavior, moreover, reflects the influence of a multitude of genes (they are "polygenic"). Genes play an important role in the cluster of causes that lie behind behavior, but are not "the" cause of behavior.[59]

When genetic influence is confused with deterministic action, it can be assured that genuine freedom to act altruistically would be called into question. However, genes do not act in isolation, nor do they have unlimited control over a person. This lack of total control over human action allows persons to be causal agents.

Taking exception to determinist ideas, Pope brings up a necessary assumption (given our natural state) where he says that if all behavior is caused by natural selection then people should not be held accountable for their actions, whether those actions are considered good or bad.[60] If we have any freedom at all, then it is in fact possible at least in principle to overcome our biological constraints to choose good actions, even

---

[56]A point made by Stephen Pope. See *Human Evolution and Christian Ethics*, 163. He seems to be attacking the subtle language of Dawkins and the gene's power over humans—as if you could separate it out from humanity.

[57]A point also made by ibid.

[58]Edward O. Wilson, *On Human Nature*, 25th anniversary ed. (Cambridge, MA: Harvard University Press, 2004). See chap. four specifically.

[59]Pope, *Human Evolution and Christian Ethics*, 163.

[60]Ibid., 161. See also Wright, *The Moral Animal*, 203.

though some people might have more constraints than others. This is the subject to which we now turn.

4.3 HUMANS ARE GENUINELY FREE AND CONSEQUENTLY RESPONSIBLE

While every person is influenced by genetic and environmental constraints that locate them on a spectrum of freedom, they are not completely determined. Any ability to overcome biological constraints on behavior, especially moral behavior, is going to necessitate human freedom in some capacity, which will function at varying levels for every individual. Accordingly, freedom cannot be a mere "self-delusion" as E. O. Wilson claims.[61] As I will show in this section, if this were the case, there is no real human autonomy. Furthermore, to claim, as Wilson does, that humans act *as if* they are free merely to sustain order in society, seems to be wanting.[62] Accordingly, Wilson's explanation of the lack of free will must reckon with mounting evidence to the contrary and is counterintuitive to what humans instinctually know to be the case.

If a person has the power to will what they do, or to refrain from action, then that action is considered free.[63] Conversely, as eighteenth-century Scottish philosopher Thomas Reid suggests, if every act of volition is a "necessary consequence of something involuntary," the person is not free at all and has not the "liberty of a moral agent" but rather, is subject to "necessity."[64] One has to have the ability to overcome *any*

---

[61]Wilson, *On Human Nature*, 71. For instance, Wilson states that there are no circumstances that come about without physical causes, including neurological circumstances.

[62]Pope, *Human Evolution and Christian Ethics*, 169.

[63]Thomas Reid, "The Liberty of Moral Agents," in *Delight in Thinking: An Introduction to Philosophy Reader*, ed. Scott C. Lowe and Steven D. Hales (New York: McGraw-Hill, 2007), 259-60. Reid argued that acts of altruism are always going to be dependent upon the volition of the human person. For our purposes, if one does not have voluntary will to overcome biological and cultural restraint, then one cannot be held responsible for either positive or negative ethical actions. Reid says, "The effect of moral liberty is, *that it is in the power of the agent to do well or ill*" (emphasis his). He later goes on to elucidate this by discussing culpability and genuine freedom, which are necessarily dependent upon each other. Thus, the question remains if humans actually have this power to overcome constraints. As it stands, when looking at the influences on the human condition in regards to altruism, the choice is not as simple as *merely choosing* between equal decisions—some choices are much more influenced than others.

[64]See also Thomas Reid, *Essays on the Powers of the Human Mind; to Which Are Added, an Essay on Quantity, and an Analysis of Aristotle's Logic* (London: T. Tegg, 1827), as well as Reid, "The Liberty of Moral Agents," 259-60.

coercion in order to be considered free. We can see demonstrations of this kind of exercise in freedom, overcoming biological and environmental constraints, with examples like human suicide, which is an extreme example of what Charles Taylor says about freedom: "One is free only to the extent that one has effectively determined oneself and the shape of one's life."[65]

If an individual does not recognize her or his ability to have human autonomy, then she or he might always be a slave to some combination of external and internal constraints. If a person cannot distinguish what they *want* to be from what they *ought* to be in order to be fully human and free—looking toward a higher self—a person could be free from cultural and biological limitations and yet not be wholly free or "self-directed."[66] This includes the ability of the mind to "self-update" itself and expand in ever changing circumstances.[67] These flexibilities allow humans to evolve and adapt in mutating environments. Consequently, what one means by "human freedom" has to be expanded to more than just "noninterference" in human affairs.[68] As noted earlier, humans are obviously influenced (or one might say, "interfered" with) by a myriad of circumstances. Rather, true liberty has "some essential connection to a substantive moral vision" where to be genuinely free is to be in command of one's life and purposes.[69]

Yet this kind of freedom brings immense responsibility to the agent. In regard to whether or not humans are free or determined, the culpability of one's actions is highly dependent upon how "determined" one is. In other words, what is the nature of that determination, and how influential is it on a particular person? William Schweiker uses an interesting metaphor regarding Adam and Eve to make his point.[70] He questions

---

[65]Charles Taylor, *Philosophy and the Human Sciences*, Philosophical Papers (Cambridge: Cambridge University Press, 1985), 213. See also Charles Taylor, "What's Wrong with Negative Liberty," in *Contemporary Political Philosophy: An Anthology*, ed. Robert E. Goodin and Philip Pettit, Blackwell Philosophy Anthologies (Malden, MA: Blackwell Publishing, 2006), 388.

[66]William Schweiker, *Responsibility and Christian Ethics*, New Studies in Christian Ethics (Cambridge: Cambridge University Press, 1995), 140.

[67]Owen J. Flanagan, *The Science of the Mind*, 2nd ed. (Cambridge, MA: MIT Press, 1991), 304.

[68]Schweiker, *Responsibility and Christian Ethics*, 140.

[69]Ibid.

[70]The following example is taken from ibid., 143.

whether the unfortunate couple should really be held blameworthy for their actions if God (or nature, from a secular perspective) *forced* them to eat the apple. Of course they are not responsible if they were determined to act, and few would hardly suggest otherwise. Neither judicial systems nor everyday life experience express situations where humans are not in some part responsible for their actions.[71]

Merely because an individual might be influenced by her or his environment or genes does not negate his or her moral freedom if he or she happens to endorse those inclinations.[72] The mind provides humans with incredible freedom from genetic "control structures" typically found in lower organisms.[73] Humans have different levels of self-understanding that unavoidably makes them moral agents, which, necessarily, relates to human responsibility.[74] As Schweiker says, "The freedom to revise one's life through the examination of values and desires is not incompatible with determining factors on our lives."[75] This freedom, however, brings with it a serious liability, which ends up condemning the individual to freedom (rather than just being "liberated" with freedom).

As we shall see in subsequent chapters, John Wesley's groups are structured to encourage both this kind of freedom and embrace this grave responsibility. Even if "absolute"[76] freedom does not exist—that is, freedom without any constraints or hindrances—it does not totally mitigate culpability of human action. Depending on the capacity of an individual to make a free choice, she or he is responsible, yet only accountable to the capability in which her or his choice was restricted. Thus, depending on where an individual resides on the spectrum of constraints (whether genetic or environmental) will *necessarily* impact the individual's

---

[71]David Eagleman has become popular of late as he discusses how neuroscience should be impacting our judicial system. He suggests that judicial systems have to account for biological impulses that might coerce people's actions. For a quick synopsis of his ideas, see David Eagleman, "Incognito: What's Hiding in the Unconscious Mind," interview with Terry Gross, *Fresh Air*, May 31, 2011 (National Public Radio, 2011).

[72]Schweiker, *Responsibility and Christian Ethics*, 146.

[73]Pope, *Human Evolution and Christian Ethics*, 174.

[74]Schweiker, *Responsibility and Christian Ethics*, 147.

[75]Ibid.

[76]Dennett makes this point to say that we have to throw away all science if individuals are not responsible or absolutely free. See Daniel C. Dennett, *Elbow Room: The Varieties of Free Will Worth Wanting* (Cambridge, MA: MIT Press, 1984), 49.

responsibility. For example, if one's moral choice is severely restrained by some kind of biology (say, a predisposition to alcoholism[77]), the individual is less responsible than someone who does not have that influence. Where one resides on the range of freedom parallels his or her responsibility. Likewise, the opposite is true if one is relatively unhampered by constraints. For instance, if another individual has no genetic proclivity to alcoholism, and was not influenced by environmental sways (such as growing up with alcoholic parents), she or he is more accountable for her or his alcoholic behavior.

In ethics, there has to be a "culpability baseline" for a person to be morally responsible. There has to be a place where an individual could have "done otherwise."[78] If this is not the case, an individual is not free (in the genuine sense of the word).[79] The volition of the agent is critical if one is to be considered morally responsible—coercion or uninhibited constraint of any kind cannot totally impede freedom and still hold individuals responsible.[80] After all, humans are not separate beings from their bodies but instead have a mind that is a product of the brain.[81] Any notion of responsibility is going to depend on the quality of freedom in a person's actions. Is the act voluntary or forced in any way? Is it merely

---

[77]As highlighted in the introduction, certain First Nations ethnic groups lack key enzymes that breakdown alcohol efficiently (after consumption). This gives them a genetic predisposition to become alcoholics. See T. Kue Young, *The Health of Native Americans: Toward a Biocultural Epidemiology* (New York: Oxford University Press, 1994), 210-11. Kent Dunnington notes in *Addiction and Virtue*: "Early advances in the genetics of addiction were the result of studies of the difference in alcoholism rates between fraternal and identical twins. In general, identical twins had more similar rates of alcoholism than did fraternal twins although there was nothing approaching an exact correspondence. In an effort to control for environmental factors, studies were then conducted on adopted children who were separated at birth from their biological parents. In general, adoptees that had at least one alcoholic biological parent were found more likely to be alcoholic than adoptees who did not have an alcoholic biological parent. In one study, the rate of alcoholism in the former group was four times that of the rate of alcoholism in the latter." See *Addiction and Virtue: Beyond the Models of Disease and Choice*, Strategic Initiatives in Evangelical Theology (Downers Grove, IL: IVP Academic, 2011), 21. See also chap. thirteen of Goodwin, *Alcoholism: The Facts*.

[78]Schweiker, *Responsibility and Christian Ethics*, 142.

[79]Again, for a thoughtful discussion of this, see Eagleman, "Incognito: What's Hiding in the Unconscious Mind."

[80]Schweiker, *Responsibility and Christian Ethics*, 142.

[81]Here, Dennett is right to criticize Descartes's dualism. He calls it "notorious" because it permeates not just the academy but also our everyday thinking. See Daniel C. Dennett, *Kinds of Minds: Toward an Understanding of Consciousness*, Science Masters (New York: Basic Books, 1996), 77.

influenced or is it totally coerced? Humanity operates under the assumption that individuals are regarded as free in most respects. Thus, humans are responsible because in some basic sense an individual controls the destiny of her or his life.[82] One can see this in the way humans operate in judicial matters, in which the key question revolves around if someone was acting freely or not.[83] As Peter French mentions, these ideas of responsibility create the "conceptual superstructure in which questions of metaphysical freedom or moral freedom are conceived."[84] As stated in his L'être et le néant, Jean-Paul Sartre's famous line, "Nous sommes condamnés à la liberté" (we are condemned to freedom), gets to the concept of freedom and responsibility.[85] This freedom, albeit varying in degrees for every individual, brings with it a very real sense of burden and accountability: our actions are our own.

To this point, J. R. Lucas brings forth a helpful statement in regard to how humans operate in the everyday: "The central core of the concept of responsibility is that I can be asked the question, 'Why did you do it?' and be obliged to give an answer. And often this is quite unproblematic. But sometimes I cannot answer, cannot be expected to answer the question 'Why did you do it?,' and then I say 'I am not responsible.'"[86] What Lucas is trying to address is the connection between constrained actions, possibly via unconscious behavior, and the common understanding within most societies concerning the placing of blame. Children often have to learn the concept of responsibility at an early age. When a younger child accidentally breaks a toy of her or his older sibling, the older child is taught to judge conscious decisions over unwitting behavior. In determining what actions to take, an individual often considers moral or ethical principles that help her or him internalize the decision.[87]

There is a clear connection between human freedom and responsibility, and if the biological constraints on an individual's behavior cannot

---

[82]Schweiker, Responsibility and Christian Ethics, 137.

[83]Peter A. French, Responsibility Matters (Lawrence: University Press of Kansas, 1992), 3.

[84]Ibid.

[85]J. P. Sartre, L'être et le néant: Essai d'ontologie phénoménologique (Paris: Gallimard, 1976), 541.

[86]J. R. Lucas, Responsibility (Oxford: Oxford University Press, 1993), 5. See also Schweiker, Responsibility and Christian Ethics, 136.

[87]Schweiker, Responsibility and Christian Ethics, 137.

be overcome then a person cannot be held accountable. As Dennett says, "But if I am nothing over and above some complex system of interactions between my body and the memes that infest it, what happens to personal responsibility? How could I be held accountable for my misdeeds, or honored for my triumphs, if I am not the captain of my vessel? Where is the autonomy I need to act with free will?"[88] Dennett is using the word *autonomy* here to mean "self-control," and it is this control over environmental and genetic influences that an agent needs to necessarily take on if they are to bear responsibility.[89]

It is true that, like all other animals, humans are extremely influenced by both genes and the environment, as we saw in chapter three, in what might be called "constraints" to an individual's behavior. It is also true that humans have a capacity to make free choices in what is considered to be free will. However, we must still question how humans overcome such biological limitations in order to perform acts of altruism. The answer is likely not as simple as claiming that an individual should merely "do good to others" because not all individuals start at the same place. On the contrary, it might even be difficult for some to do the opposite ("do bad to others") in certain situations (depending on their constraints). If humans in fact have some semblance of free will, typically making decisions using intelligence, and do not unswervingly act on genetic impulse to behavior,[90] then rebelling against one's genes, as Dawkins and Dennett would have it, is the goal if one wants to overcome genetic and environmental constraints. This is especially the case if one wants to act in altruistic ways, caring for the welfare of others as an end in itself, rather than merely looking out for one's own gene fitness.[91]

---

[88]Daniel C. Dennett, *Darwin's Dangerous Idea: Evolution and the Meanings of Life* (New York: Simon & Schuster, 1995), 366.

[89]Although it is not totally relevant for the current discussion, an interesting point is made by Schweiker in regard to Christian tradition of culpability: "Thinkers like Augustine, Calvin, and others have argued that agents are responsible for their character and conduct even though they were not able to act otherwise than in sin. These claims are strictly theological in character; they concern the status of the agent *before God*. And given this, only God can redeem fallen sinful humanity; human beings cannot merit grace through their actions." He goes on to say, "Yet this argument does not obviate the *moral* insight. In order to be held responsible an agent must act voluntarily even if human action is circumscribed within a religious vision of reality." See Schweiker, *Responsibility and Christian Ethics*, 142.

[90]Pope, *Human Evolution and Christian Ethics*, 174.

[91]Elliott Sober and David Sloan Wilson, *Unto Others: The Evolution and Psychology of Unselfish*

For humans, then, the ability to implement moral responsibility is made possible by the evolving human brain and developed cognitive and emotive faculties.[92] Human consciousness plays a powerful role in giving the human the capacity to step beyond her or his own self and take on a new perspective.[93] Though scientists can locate definite brain patterns involved in particular actions, it is the *whole person* who is conscious, who decides and who acts.[94] Whether endowed by God or developed as the product of some other kind of process, humans have a propensity toward creating their own destiny while being strongly influenced by their biological history. Dennett, arguing from a naturalist perspective, makes the case that although genuine human freedom exists, it is not some kind of "God-like" power that makes humans different from other animals, nor is it some "pre-existing" state that has always been.[95] Rather, for Dennett, it is an evolved product of human activity (like other aspects of culture or environment).[96]

Although evolution does play a role in human behavior, of course, humans seem to be the only animal that struggle against biology to make certain moral decisions. According to Dennett, a human goes through a two-step process of making decisions. When faced with a decision to act, humans have a "consideration-generator"[97] that floods the person with options of potential actions. Next, the individual sifts through the plausible decisions and outcomes, which become predictive of the person's final decision.[98] Consequently, the person works through the biological in order to work out the best possible decision.[99] Yet, Dennett also holds

---

Behavior (Cambridge, MA: Harvard University Press, 1998), 228.

[92]Pope, *Human Evolution and Christian Ethics*, 185.

[93]This perspective that often takes an emotional angle that impacts decision making.

[94]Pope, *Human Evolution and Christian Ethics*, 185-86.

[95]Dennett, *Freedom Evolves*, 13.

[96]To this qualified concept of human liberty, he attributes freedom, akin to music and money, which is a product of environmental influences. See ibid.

[97]This phrase is interesting for Dennett. It seems to possess some kind of suprascientific ability (or at least not solely a biological ability) to conjure up these series of considerations. It also suggests a Cartesian dualism of the mind/brain, a concept not totally compatible with Dennett's ideas. In general, materialist naturalism is dependent on Cartesian dualism because it assumes that matter is what Descartes said it was, and simply denies that mind exists as a separate substance.

[98]Daniel C. Dennett, *Brainstorms: Philosophical Essays on Mind and Psychology* (Montgomery, VT: Bradford Books, 1978), 295.

[99]In this way, strictly speaking, Dennett is a compatibilist because there is some free will, but no freedom outside the biological constraints.

that even though humans are determined by their biological composition, and by this I mean to say that they are dwelling within parameters
of constraints that are determined, humans have a distinct place in
moving within these parameters, shaping part of their own destiny.
Dennett says,

> Whales roam the oceans, birds soar blithely overhead, and, according to an
> old joke, a 500-pound gorilla sits wherever it wants. But none of these crea-
> tures is free in the way human beings can be free. Human freedom is not an
> illusion; it is an objective phenomenon, distinct from all other biological con-
> ditions and found in only one species, us. The differences between auton-
> omous human agents and the other assemblages of nature are visible not just
> from an anthropocentric perspective but also from the most objective stand-
> points (the plural is important) achievable. Human freedom is real—as real
> as language, music, and money—so it can be studied objectively from a no-
> nonsense, scientific point of view.[100]

This is the human freedom that individuals engage to rise above the
"tyranny of the genes."[101] This compatibilist freedom that humans have
can be extended to matters of altruism.

While humans have the capacity to make such moral decisions, animals also have a kind of freedom. To show the unique ability of humans
to willfully overcome constraints on moral behavior, it might be helpful
to briefly compare humans to other animals. Dogs provide a clear example of such differences in freedom. When one calls a dog to go outside,
sometimes the animal wavers back and forth and seems to be making a
decision. Yet, according to Pope, in comparison to dogs, evolution has
granted humans an "enhanced capacity for voluntary action and a capacity to select what we think are the best means to the end that we
identify as good for us."[102] The difference is stark: the dog is concerned
with ideas of immediate implications and base urges, whereas the human
often makes decisions with a focus on what is good. Human beings are
inclined to seek toward some good—the main challenge revolves around

---

[100]Dennett, *Freedom Evolves*, 304-5.
[101]See Dennett, *Darwin's Dangerous Idea*, 367. Here he is making a comment about Dawkins famous line about the "tyranny of the selfish gene." See Dawkins, *The Selfish Gene*, 200-201.
[102]Pope, *Human Evolution and Christian Ethics*, 178.

what actually constitutes the "good-life."[103] As previously mentioned, even according to Dawkins, humans have the capacity to "defy the selfish genes of our birth"[104] and overcome our biology. Humans, then, have the distinct ability to consciously look out for long-term advantages over short-term selfishness.[105] They can also, according to Dawkins, "discuss ways of deliberately cultivating and nurturing pure, disinterested altruism—something that has no place in nature, something that has never existed before in the whole history of the world."[106]

Humans, then, despite common ancestry, function very differently than most animals. Charles Darwin himself noted this. Turning the focus to the human individual, he says in *The Decent of Man* that if humans patterned themselves after bee social behavior it would be expected that in a similar fashion to worker bees, the unmarried human females would find it their duty to kill their brothers and "mothers would strive to kill their fertile daughters; and no one would think of interfering."[107] Darwin, however, was not trying to prove that "might is right" like Thrasymachus wanted to argue in Plato's *Republic*.[108] Rather, Darwin quipped, "I have noted in a Manchester newspaper a rather good squib, showing that I have proved 'might is right' and therefore Napoleon is right and every cheating tradesman is also right."[109] Humans have the ability to act on some influences and not others. Employing those freedoms depends on the capacity one has for free choice (overcoming constraints).[110] If "selfishness" is being concerned with one's own interests,[111] then it is safe to say that humans will have to overcome some

---

[103]In Aristotle, *Nicomachean Ethics*, trans. Martin Ostwald, The Library of Liberal Arts 75 (Indianapolis: Bobbs-Merrill, 1962), 1094aI. Found in Pope, *Human Evolution and Christian Ethics*, 178-79.

[104]Dawkins, *The Selfish Gene*, 200.

[105]Pope, *Human Evolution and Christian Ethics*, 166.

[106]Dawkins, *The Selfish Gene*, 201.

[107]Charles Darwin, *The Descent of Man, and Selection in Relation to Sex*, 2 vols. (New York: D. Appleton and Company, 1871), 70.

[108]This idea was discussed well by Michael Ruse at the 2011 Ian Ramsey Center Conference called "The Evolution of Morality and the Morality of Evolution."

[109]See James Rachels, *Created from Animals: The Moral Implications of Darwinism* (Oxford: Oxford University Press, 1990), 62. See also Wright, *The Moral Animal*, 330. See also Charles Darwin and Mark Ridley, *The Darwin Reader* (New York: Norton, 1987).

[110]Pope, *Human Evolution and Christian Ethics*, 183.

[111]Ayn Rand and Nathaniel Branden, *The Virtue of Selfishness: A New Concept of Egoism* (New York: New American Library, 1965), x.

of those interests in order to be altruistic to others. But unlike other animals, humans are in a prime position to determine their future because they have the broadest knowledge and thus the best perspective on the future.[112] Pope notes,

> In a Christian ethical perspective, *the* human responsibility is to engage, as deeply and seriously as possible, in lives of ethical "sublation": to employ the capacities given to us by God through the evolutionary process for the highest ends for which we can strive, to use the capacity for free choice in ways that move us to closer union with the ultimate good, and to live in such a way that we are increasingly drawn to goodness and virtue. From the perspective of Christian ethics, evolution is the means employed by the Creator to produce creatures, including human beings, and human freedom is best understood in terms of evolved human nature and not as a bizarre exception to everything else known scientifically about human beings. Whereas the capacities that make us distinctively human emerged through the blind working of the evolutionary process, virtues never appear without the active cooperation of human agents.[113]

Humans are agents capable of steering the course of their lives.[114]

It is to this point that Pope is helpful in bringing out that humans are not necessarily above other animals, but rather are significantly more advanced. In this way, he helps one avoid a kind of "special creation" and "God of the gaps" argument, when he states,

> One could argue that the consciousness of different species of animal also runs along a spectrum, and that, again, the fact that we reside on one end of the spectrum does not imply that we are the only ones on it. If many higher animals have consciousness, and some even self-consciousness, there seems no reason to postulate a special divine initiative to account for this feature of human nature or to think that God has to provide any missing power that is not already a potency of nature itself.[115]

---

[112]Robert H. Frank, *Passions Within Reason: The Strategic Role of the Emotions* (New York: Norton, 1988), 235.

[113]Pope, *Human Evolution and Christian Ethics*, 186-87.

[114]Schweiker, *Responsibility and Christian Ethics*, 145. This is why the connection between freedom and the assignment of responsibility is critical for any moral justification or punishment, or any culpability for one's actions. See also ibid., 158.

[115]Pope, *Human Evolution and Christian Ethics*, 175.

The characteristics of free choice are not exclusive to humans, yet humankind does exercise unprecedented freedom compared to other animals. Christians tend to have a plethora of "God of the gaps" arguments even today,[116] and it does not seem necessary to make a claim that because of humankind's great freedom, freedom itself must have divine and not natural origins. But regardless of where freedom originates, God or nature, this amalgam of consciousness and freedom empowers humans to overcome biological urges to *only* care about gene fitness and instead to consider what is good. Merely because one recognizes that certain evolved attributes play a definitive part in our responses to experience does not necessitate that an individual's whole character is "nothing but an epiphenomenon of his or her genes."[117] Rather, moral freedom reflects the view of character that is developed through the routine choice of acts over which the agent has some control.[118] As Pope mentions, even the phrase *human nature* conveys the idea that humans are "*inclined* to respond to what they perceive to be objects that fulfill some kind of essential need. But it does not, in and of itself, mean that this response is automatically preordained by genes or the evolutionary process."[119] In a way, the origin of this kind of rebellion against the genes does not change the reality that humans are "specially" composed to do so.

Even Wilson also suggests that humans still have the power to follow some urges of human nature and to overpower others.[120] What actually distinguishes human freedom from other animal freedom is a person's ability to act *in spite* of many inner divisions of influences.[121] To think that humans are influenced by biology and Dennett's "consideration-generator" and yet still come away making decidedly different choices from one another exemplifies the ability of a person to rise above constraints to choose a certain action. Human action, therefore, is not entirely undetermined—it is not wide open with no external influences.

---

[116]James Edward Huchingson, *Religion and the Natural Sciences: The Range of Engagement* (Fort Worth, TX: Harcourt Brace Jovanovich College Publishers, 1993), 135.

[117]Pope, *Human Evolution and Christian Ethics*, 185.

[118]See ibid. See also Aristotle, *Nicomachean Ethics*, 1114b.

[119]Pope, *Human Evolution and Christian Ethics*, 180. Emphasis his.

[120]Wilson, *Consilience*, 97.

[121]Mary Midgley, *The Ethical Primate: Humans, Freedom, and Morality* (London: Routledge, 1994), 7.

Rather, a free choice for an individual is "neither necessitated nor re-strained by external coercion" despite actions made by routine or habit.[122] Each individual might have to overcome greater or lesser constraints in order to get to a place where her or his actions are freely chosen. Yet regardless of where individuals fall on that spectrum, humans have this "free" capability. Humans are not slaves to their genes, they are free to overcome and ignore them.[123]

Dennett posits a helpful clarification to the idea of overcoming con-straints: the Kantian idea of developing one's reason so one's emotions do not overcome it when it is unwarranted.[124] Here, Dennett says,

> I cannot micromanage my own real-time deliberations, so I have to resort to shotgun approaches, equipping myself with powerful emotional dispositions that spill over their targets, leave me trembling with rage when rage is ap-propriate, unable to contain my joy when joy is appropriate, swept away by sorrow or pity. But in order to get these emotions to help me make long-term prudential decisions when I face temptation from short-term Sirens, I have to let them rule me as well when my choice is between my short-term gain and what is best for others.[125]

Thus, there is a constant give and take with how much a person should override innate urges. After all, these natural inclinations are what brought the individual through eons of evolution. Yet still, the key to gaining a reputation for being good—which can work advantageously for gene fitness in the long run—is actually *being* good.[126] One will always fight the urge to battle short-term selfishness (which provides quick gain) against long-term rewards of cleverly timed selflessness.

For example, Neil Messer explores how humans overcome behavior that is constrained by such powerful evolutionary forces. Drawing from

---

[122]Pope, *Human Evolution and Christian Ethics*, 182. Here, Christian ethics might prove helpful. A person's own choice is enough to generate a free action and is made achievable by both the evolutionary construction of the person, and by the person being made in the image of God.

[123]Ridley, *The Red Queen*, 16.

[124]See Immanuel Kant, *The Critique of Judgement*, trans. James Creed Meredith, 2 vols. (Oxford: Clarendon Press, 1952). And see also Kant, *Immanuel Kant's Critique of Pure Reason*, trans. Norman Kemp Smith (London: Macmillan, 1929).

[125]Dennett, *Freedom Evolves*, 214.

[126]Dennett actually says that there are no "short-cuts" to being good, yet quips that "evolution is still going on." See ibid.

Richard Wrangham and Dale Peterson's research on male violence and sexual aggression in bonobos,[127] Messer concludes that while human men have similar propensities that connect them to their primate cousins, traits such as these are not completely determined. He says,

> The lesson Wrangham and Peterson draw for human life from the comparison of chimpanzees and bonobos is that men are likely to have a deep-seated propensity to status-seeking pride and groupish aggression for the foreseeable future, but that the more social power and influence is exercised by women, and the stronger the female coalitions that are formed in human societies, the more these proud and aggressive male tendencies can be curbed.[128]

One should keep in mind that, while organisms are predisposed to certain behavior, a compatibilist understanding of free will argues that all human events are caused not only by biology or the environment (or some combination of both), but also by humans. This goes against a hard determinist viewpoint that leaves off the clause about human purposeful action. For instance, parents of multiple children, one of whom is unlikely to reproduce (due to disease or disability), may have the biological urge to neglect that child and afford resources to the others. According to Robert Trivers, this urge likely came about through natural selection that favored parents who used limited resources to help the children with high reproductive fitness.[129] Biologist Janet Mann tested this theory on twins of low birth weight, albeit with a small sample size, and found that mothers tended to favor the healthier twin after eight months.[130] Yet, there is more to the story than Mann tells in this study. One has to take into account the whole person. Without a doubt, humans are dramatically impacted by such proclivities from natural selection, as Trivers and Mann suggest. However, there are numerous cases in human history

---

[127]See Richard W. Wrangham and Dale Peterson, *Demonic Males: Apes and the Origins of Human Violence* (Boston: Houghton Mifflin, 1996).

[128]Neil Messer, *Selfish Genes and Christian Ethics: Theological and Ethical Reflections on Evolutionary Biology* (London: SCM, 2007), 136.

[129]Robert L. Trivers, "Parental Investment and Sexual Selection," in *Sexual Selection and the Descent of Man, 1871-1971*, ed. Bernard Grant Campbell (Chicago: Aldine, 1972).

[130]Janet Mann, "Nurturance or Negligence: Maternal Psychology and Behavioral Preference Among Preterm Twins," in *The Adapted Mind: Evolutionary Psychology and the Generation of Culture*, ed. Jerome H. Barkow, Leda Cosmides and John Tooby (New York: Oxford University Press, 1992).

where, despite such genetic compulsions, parents go so far as choosing to *adopt* disabled children of little reproductive fitness. It seems what Mann proved is merely the idea that humans are prone to certain behavior, not unilaterally controlled by biological urges. As Messer points out, we do not have to deny the fact that humans have predispositions given to them by evolutionary history, and that in some cases, these proclivities hinder humans from behaving altruistically;[131] but these tendencies in and of themselves are not deterministic mechanisms that write the fate of human actions.

To this point, Ridley says that humans operate with the assumption that individuals have some control over their own behavior and will not purposefully take advantage of others. Or, at least, not make a lifestyle of doing so. He says, "We trust strangers, tip waiters we will never see again, give blood, obey rules and generally cooperate with people from whom we can rarely expect reciprocal favours. To be a selfish free-rider is such a sensible and successful strategy in a large group of reciprocating cooperators . . . that it seems crazy more people do not choose such an option."[132] Crazy indeed. What is interesting about this idea is that people could readily profit from others by carefully taking advantage of those around them. This practice certainly happens, to be sure, but not to the extent one would think if one is to believe that all human behavior is generated from some kind of selfish gene that merely cares about begetting to the next generation.[133]

Before proceeding, I will provide a short aside that is pertinent to what I will be discussing in the chapters to come. The Christian tradition has not been interested in the freedom of the will as a separable part of the human person, but rather in the whole person: freedom, environment and biology together. This is to say, the free will that directs a human is derived from a sense of doing what is good (for the *whole* person)—not

---

[131]Messer, *Selfish Genes and Christian Ethics*, 145-62.

[132]Ridley, *The Origins of Virtue*, 180.

[133]In doing research in sociobiology, I am often surprised that many sociobiologists—such as Dawkins—do not take their ideas out to the logical conclusion. If what matters is helping genetic code move forward then why not volunteer services at a fertility clinic, or sperm bank, or live a highly promiscuous life? This would without a doubt be the fastest way to achieve this goal. Instead, there is lip service to morals, but these ideas are often barely supported. At least not supported to the extent they should be.

merely good for their genes, finances or any other parsed-out bit of human existence.[134] Sociobiologists, on the other hand, tend to describe free will as part of a person, to the extent that one wonders if they are not describing a moral organ, which would imply a kind of dualism. Yet in reality, there is no single controlling part of the brain just as there is no "unified governing psychic center" that enslaves the rest of the body for its own purposes.[135] One might say, as Peacocke so elegantly expresses in *God and the New Biology*, "God has made human beings thus with their genetically constrained behavior—but, through the freedom God has allowed to evolve in such creatures, he has also opened up new possibilities of self-fulfillment, creativity, and openness to the future that requires a language other than of genetics to elaborate and express."[136] In much the same manner, Pope says, "Rather than constituting a dichotomy, then, the practice of human responsibility shows that our evolved human nature can be actualized in a way that radiates freedom in its most profound sense."[137]

Even though humans are intimately connected to the natural world by sharing common ancestry, humanity is also distinguished from other animals. Like them, humans have natural tendencies to take care of basic needs (such as food, security, reproduction); but in addition, humans have inclinations to do what is good in general or what is "comprehensively good."[138] This is a claim that Dennett also repeatedly makes, calling humans "set apart" from other animals in regard to freedom.[139] Surprisingly, Dawkins concedes to this as well. In his *The Extended Phenotype*, he discusses the inability of genes to control the whole human. He also mentions how humans can overcome both biological and environmental urges by saying, "Genetic causes and environmental causes are in principle no different from each other. Some influences of both types may be hard to reverse; others may be easy to reverse. Some may be usually hard

---

[134]This concept is fleshed out significantly in Pope, *Human Evolution and Christian Ethics*, 180-81.
[135]This is similar to Descartes thinking that he located the human soul (in the pituitary gland of the brain). See ibid., 181.
[136]A. R. Peacocke, *God and the New Biology* (San Francisco: Harper & Row, 1986), 110-11.
[137]Pope, *Human Evolution and Christian Ethics*, 187.
[138]Frank, *Passions Within Reason*, 91.
[139]He uses the words *alone* from other animals. See Dennett, *Darwin's Dangerous Idea*, 365.

to reverse but easy if the right agent is applied. The important point is that there is no general reason for expecting genetic influences to be any more irrevocable than environmental ones."[140]

Dennett is correct in saying that humans are somehow different than other animals in their capacity to live freely. Yet human freedom, stemming from natural causes, is constrained by the genetic and environmental history of the individual. Those influences contain the parameters of freedom but do not fully hamper authentic choice. Each person differs in where they dwell on the continuum of constraints, which can be located in one's family of origin, as well as one's communities.[141] As we shall see in the following chapters, John Wesley intuitively picks up on this human condition and establishes a system of environmental constraints on his followers to encourage altruistic action.

4.4 CONCLUSION

In this chapter, I hoped to demonstrate how human behavior is influenced by genetic and environmental constraints, but not determined. These restrictions necessarily impact human freedom and, subsequently, responsibility. However, because humans are not wholly subjugated to these influences, it is within their power to overcome such constraints and practice altruism. Nevertheless, not all individuals start at the same location on the selfish/selfless spectrum. Certain people who have stronger genetic drives toward selfishness will have a more difficult time overcoming those urges in order to perform altruistic actions. Others with less biological complications, or those with environmental influences that encourage altruistic behavior, are more able to make altruistic decisions. It is important to note that genes alone cannot explain the totality of human behavior.[142] Yet, in regard to the composition of

---

[140]Richard Dawkins, *The Extended Phenotype: The Gene as the Unit of Selection* (Oxford: Freeman, 1982), 13. Regarding phenotypes and environmental epigenetics, in Eva Jablonka, Marion J. Lamb and Anna Zeligowski's book, called *Evolution in Four Dimensions*, they show through an experiment done on the varying the color of mouse fur that "environmental and developmental factors can affect the expression of a gene and lead to variant phenotypes." For the full experiment see Jablonka, Lamb and Zeligowski, *Evolution in Four Dimensions: Genetic, Epigenetic, Behavioral, and Symbolic Variation in the History of Life*, rev. ed. (Cambridge, MA: MIT Press, 2014), 140.
[141]Pope, *Human Evolution and Christian Ethics*, 183.
[142]Ibid., 163.

humankind, it is also unnecessary to appeal to a "God of the gaps" argument. Pope says,

> The universe is organized in such a way as to generate life, and then to generate more and more complex forms of life, and then to generate more and more intelligent forms of life. God sustains the creation in being but does not need to guide the evolutionary process. This is the "fully gifted" cosmology. . . . God uses the structures of nature and contingent events to produce creatures, among whom are human beings. In this view God can interact with nature in such a way as to guide its unfolding, but need not do so. There is no need to refer to supernatural agency in an explanation of natural history, and no need for a "God of the gaps."[143]

There is no need to believe that God cannot work within the natural world in order to bring about good things. Consequently, in the following chapter we will see how John Wesley's theology makes room for a God to work in this way.

---

[143]Ibid., 176; Howard J. Van Till, *Portraits of Creation: Biblical and Scientific Perspectives on the World's Formation* (Grand Rapids: Eerdmans, 1990).

5

WESLEYAN HOLINESS AGAINST
A BACKDROP OF EVOLUTION

◆

Due to the fact that these selfish and selfless tendencies are so tied to human evolution, we must address the question of the compatibility of sociobiology with Wesleyan ethics. It is thus the intent of this chapter to describe how one might understand Wesleyan holiness against a background of evolutionary biology, demonstrating how—by focusing on Wesley's theological assumptions, which are regarded as the foundation for Wesleyan ethics—sociobiology and Wesleyan ethics are, in fact, compatible. In this way, this current chapter will serve as the framework for the final chapter, where I will demonstrate the ways in which John Wesley built upon the biological makeup of his followers and developed highly structured groups that mitigated selfish behavior while nurturing altruistic inclinations.

To accomplish such a theological foundation, we must first appeal to Wesley's quest for holiness. Wesley was enamored with the concepts of holiness and sanctification, creating and structuring his small groups around such theological concepts. Though I will explore how Wesley used holiness as the keystone of his movement, I will also elaborate how his concept of holiness shed light onto his ideas concerning the human condition. Within this discussion, I will pursue his theological underpinnings and his nuanced perspective of original sin and Christian perfection. Last, and most significantly, I will put forth a case that Christian perfection—one of the most critical concepts in Wesleyan ethics—is compatible with our current knowledge of evolutionary biology. What is more, though Wesley's world of constraints is open for critique with regard to the notion of perfection, I will defend his views by redefining how we might understand Christian perfection in light of what we now know about human biology. (These topics will be discussed further in chapter six.)

Wesley's concept of Christian perfection, also known as entire sanctification or holiness, is perhaps one of the most misunderstood doctrines in Wesleyanism for both nonfollowers and followers of Wesley alike. This misunderstanding has its roots in Wesley's disapproval of the idea that one cannot achieve perfection in this life.[1] Christian perfection in Wesleyanism concerns a robust doctrine of spiritual transformation;[2] as Michael Christensen defines it, Christian perfection is "an experience of grace, subsequent to salvation, with the effect that the Holy Spirit takes full possession of the soul, sanctifies the heart, and empowers the will so that one can love God and others blamelessly in this life."[3] Such a doctrine of Christian perfection helps us, then, to understand that God grants grace to his followers, yet does not *replace* biological constraints on our behavior. In "A Plain Account of Christian Perfection," Wesley notes that

> it may be observed, this sermon was composed the first of all my writings which have been published. This was the view of religion I then had, which even then I scrupled not to term *perfection*. This is the view I have of it now, without any material addition or diminution. And what is there here, which any man of understanding, who believes the Bible, can object to? What can he deny, without flatly contradicting the Scripture? What retrench, without taking from the word of God?[4]

Here we find Wesley admitting that, despite the difficulty of such a concept, he remained unwavering in his thoughts concerning Christian perfection his entire life. Wesley wanted to make clear that he believed

---

[1]Marselle Moore, "Development in Wesley's Thought on Sanctification and Perfection," *Wesley Theological Journal* 20, no. 2 (1985): 29.

[2]Michael J. Christensen, "Theosis and Sanctification: John Wesley's Reformulation of a Patristic Doctrine," *Wesley Theological Journal* 31, no. 2 (1996): 71.

[3]Ibid.

[4]John Wesley, *The Works of John Wesley*, ed. Thomas Jackson and Albert C. Outler, 14 vols. (Grand Rapids: Zondervan, 1958), 11:369. Wesley's emphasis.

Christian perfection to be a biblical notion that should appear common-sense to most readers of Scripture. Consequently, for Wesley there is no need to limit the doctrine of perfection merely to "the life to come"; instead, Christian perfection is for this life and can be achieved despite any biological proclivities toward the contrary.

5.1 THE QUEST FOR HOLINESS

John Wesley obviously knew nothing of human evolution or the recent developments of Frans de Waal and other sociobiologists who claim that humans are not as naturally selfish as previously thought. Nor did Wesley know about biological tendencies toward selfishness or other compatibilist concepts as identified by Daniel Dennett, among others. Rather, Wesley's motivations for creating a world of constraints that helped to create an environment that nurtured altruism must be understood as the outcropping of theological motivations—specifically, the quest for holiness. Wesley did not see human nature in a positive light; yet he continually saw God working with individuals to help correct their natural tendencies. This transformation was a foundational concept for Wesley as he sought to explore the implications of holiness, all the while structuring a fledgling movement.

Wesley believed that a consequence of the fall for human nature, a topic I will turn to in more detail later in this section, was that all of humanity had lost the ability to completely know themselves and God. Though such a reading of the fall often leads to the doctrine that holds humans to be wholly divorced from God, Wesley had a nuanced view of total depravity. He thought that no person was completely removed from God because every human, even the non-Christian, has been favored by prevenient grace (God working in the lives of those who do not know him). As William Abraham states in *Aldersgate and Athens*, "Without radical divine assistance, we are malfunctioning cognitive agents in spiritual matters. Our rebellion against God, originating in Adam and handed down through the generations by way of original sin, has left us deeply disordered intellectually."[5] Again, this is the state of humanity *apart* from

---

[5]William J. Abraham, *Aldersgate and Athens: John Wesley and the Foundations of Christian Belief* (Waco, TX: Baylor University Press, 2010), 25. Abraham follows this section up with a quotation

the prevenient grace of God. Fortunately, Wesley had a full understanding of God's work in all humankind. Abraham notes that through prevenient grace God "universally restores in us the initial capacity to perceive the truth."[6] Thus, all persons are not completely bent toward total depravity, because all persons are recipients of prevenient grace.

For Wesley, the process of becoming holy included a nuanced concept of altruism that was tied to generosity.[7] In order to claim such, he had to create a system of practicing holiness of which the bands and classes were part (as noted in chapter six). In fact, so prominent was Wesley's emphasis on his concept of holiness that the following phrase was often (mistakenly) attributed to him: "I mean by holiness, nothing else but God stamped and printed upon the soul."[8] Wesley saw the mission of his early groups to be to live this imprinting out practically. In a sense, the term *perfection* became another way of describing "holiness." Outler mentions, "Every one that is perfect is holy, and every one that is holy is, in the Scriptural sense, perfect."[9] For Wesley, Christian perfection always consisted of wholly loving God with one's heart, soul, mind and strength, and loving one's neighbor as oneself. This view of perfection defined how Wesley understood holiness and how he transmitted it to the early Methodists. In the sermon "The Scripture Way of Salvation" Wesley says that Christian perfection is "love excluding sin; love filling

---

from Wesley's sermon 95, "On the Education of Children." Here Wesley states, "After all that has been so plausibly written concerning 'the innate idea of God'; after all that have been said of its being common to all men, in all ages and nations; it does not appear, that man has naturally any more idea of God than any of the beasts of the field; he has no knowledge of God at all; no fear of God at all; neither is God in all his thoughts. Whatever change may afterwards be wrought, (whether by the grace of God or by his own reflection, or by education) he is, by nature, a mere Atheist." John Wesley, *Sermons 71–114*, ed. Albert C. Outler, The Bicentennial Edition of the Works of John Wesley (Nashville: Abingdon Press, 1986), 350.

[6] Abraham, *Aldersgate and Athens*, 26.

[7] It is important to note that Wesley would have been unfamiliar with the term *altruism*, as it was not coined until the nineteenth century. Yet, he would have been quite familiar with the concept itself. Consequently, I use the word *altruism* in reference to Wesley with the full recognition that it may seem semantically anachronistic.

[8] This statement is instead from Ralph Cudworth, English philosopher from the seventeenth century. Yet there is no coincidence that it often gets attributed to Wesley. See Ralph Cudworth, *The Life of Christ the Pith and Kernel of All Religion: A Sermon Preached Before the Honourable House of Commons, at Westminster, March 31, 1647* (Westminster: BiblioBazaar, 2010).

[9] John Wesley, *John Wesley: A Representative Collection of His Writings*, ed. Albert C. Outler, A Library of Protestant Thought (New York: Oxford University Press, 1964), 254.

the heart, taking up the whole capacity of the soul."[10] Additionally, he wrote a sermon entirely dedicated to the concept: sermon 40, "Christian Perfection."[11] Here, salvation was the total renewal of the deformed *imago Dei* within humans.

It is important to note that the formation of Wesley's theology was a continual and robust process. As Marselle Moore has noted, Wesley's theology was, indeed, dynamic; Wesley was never static in his thinking—even up to his death—and was influenced by a variety of sources beyond the Church of England, ranging from Catholic, to Eastern Orthodox, Baptist, Quaker, Moravian and Lutheran.[12] The theological ideas of these traditions helped shape the groups, becoming, in fact, their very motivation.[13] Mildred Bangs Wynkoop, in her book *A Theology of Love*, articulates Wesley's navigation through various traditions succinctly:

> He worked out from a "system" which in his mind was not materially different from traditional Christian doctrine. He added a spiritual dimension which put theology into a new framework—personal relationship and experience. This "addition" threw the balance of doctrines into a different configuration but did not actually alter the system. His entire ministry was an explication of the altered configuration. Love, the essence of the new perspective, served as a unifying factor in theology and a humanizing application to life. The structure of theology was, under Wesley's hand, made to fit human possibilities. This does not destroy theology but it does ask penetrating questions of it.[14]

Unified in love, armed with useful aspects from various traditions, and rooted in his own Anglican community, Wesley slowly developed the

---

[10]John Wesley, *Sermons 34–70*, ed. Albert C. Oulter, The Bicentennial Edition of the Works of John Wesley (Nashville: Abingdon Press, 1985), 155-69.

[11]Ibid., 97-121.

[12]See Moore, "Development in Wesley's Thought on Sanctification and Perfection."

[13]As Wesley was pulling such concepts from various groups for pragmatic reasons, Albert Outler rightly notes that this solidified Wesley's theology. Outler states, "For Wesley, however, it was just this double notion of sin as reducible and of faith as a risky business that reinforced his stress on Christian self-discipline (moral *and* spiritual). For as the believer learns to repent daily, and to trust God's grace, and to grow in that grace, then he begins to move from the threshold of faith (justification) toward its fullness (sanctification). This particular linkage between *sola fide* (justification) and 'holy living' (sanctification) has no precedent, to my knowledge, anywhere in classical Protestantism." See Albert C. Outler, *Theology in the Wesleyan Spirit* (Nashville: Tidings, 1975), 39.

[14]Mildred Bangs Wynkoop, *A Theology of Love: The Dynamic of Wesleyanism* (Kansas City, MO: Beacon Hill Press, 1972), 19.

theological underpinnings he would need to properly construct his groups into an active and lasting movement. This movement, founded on theological concepts of holiness, was powerful enough to transform not only individual lives but also Wesley's contemporary and subsequent followers.[15]

What also set Wesley apart from his contemporaries was that he not only had an essentially catholic understanding of sin as a "malignant disease" rather than as simply a distortion of the *imago Dei* in fallen human nature, but he also displaced the popular doctrine of "election" with his own understanding of "prevenient grace."[16] This perspective allowed him to see that justification and sanctification were two facets of a single truth, separated neither by time nor experience but joined in intimate union with one another.[17] To this end, one of Wesley's main goals was to try to help people be restored to the *imago Dei* and be filled with the fullness of God.[18]

For the early Methodists, salvation included the *whole person;*[19] sin, then, was a mark on the whole person, life was more than merely individual actions and an individual's whole life was an expression of the total person.[20] Thus, Wesley did away with the Platonic dualism of the body/soul, which helped the early Methodists see that being an altruistic person was a virtue that impacted the whole person. As a result, this movement away from a Greek worldview and toward a Hebraic one was

---

[15]Paul Wesley Chilcote, *Recapturing the Wesleys' Vision: An Introduction to the Faith of John and Charles Wesley* (Downers Grove, IL: InterVarsity Press, 2004), 45.

[16]Outler, *Theology in the Wesleyan Spirit*, 34.

[17]Wynkoop, *A Theology of Love*, 20.

[18]Darrell Moore, "Classical Wesleyanism" (unpublished paper, 2011), 3. One should also not forget that Wesley had a propensity to pursue personal, continual spiritual growth. This enabled him to motivate those around him, spurring them into deeper religious practices (which necessarily fed into his culture of constraints). L. O. Hyson rightly states, "Process, growth, maturity, perfection, completeness, are words expressing key motivations for his life. He could change his theology, his politics, his ecclesiology, if he became convinced that his positions contradicted Scripture, practice, or observed reality." L. O. Hynson, W. Kostlevy and Albert C. Outler, *The Wesleyan Revival: John Wesley's Ethics for Church and State* (Salem, OH: Schmul Publishing Company, 1999), 21. Wesley was constantly striving to clarify his message and to communicate it to the people of his day. See also Moore, "Classical Wesleyanism," 11.

[19]Wynkoop actually goes further than this by mentioning that "the body is not sin-bearing; it is basically good. Sin is an attitude and spirit of rebellion against God, not a substance." Wynkoop, *A Theology of Love*, 49.

[20]Moore, "Classical Wesleyanism," 15.

pivotal for Wesley's fledgling group. In particular, one of the dangers of dualism within Methodism was making life with God less dependent on the *process* of salvation (or the many conversions along the way) and more dependent on a one-time "crisis of salvation"—a process wherein the convert is oftentimes less motivated to develop the deep practice of holiness. Surely, with any conversion, there is always a moment when one becomes converted (even if that moment is indistinguishable); for the early Methodists, however, working through the process of salvation was essential to one's deeper movement into holiness.[21] It is in this idea of the many conversions of an individual that we find Wesley's concept of "perfection" fleshed out as τελείωσις (perfecting perfection) instead of *perfectus* (perfected perfection).[22] The Wesleyan concept of *who* a person is, then, stems from Hebraic tradition, rather than Greek tradition.[23] For this paring down of his theology into a short three-position list—original sin, justification by faith alone and holiness of the heart and life—Wesley received criticism, both from the Church of England and the English evangelicals.[24] He was also criticized for drawing together the concepts of inward holiness and outward action, as his letters to George Downing on April 6, 1761,[25] and "Various Clergymen" on April 19, 1764,[26] describe.

5.1.1 Wesley's understanding of original sin

It should be clearly stated that Wesley's view of human nature is much more pessimistic than that of sociobiology. While sociobiologists previously thought humans were entirely selfish creatures, they now look at humans as being *neither* entirely selfish *nor* entirely selfless; Wesley, on the other hand, saw humanity as totally depraved creatures influenced by prevenient grace.[27] Wesley's sermon 44, "Original Sin," points

---

[21]See also ibid., 16.

[22]John Wesley, *Sermons 1–33*, ed. Albert C. Outler, The Bicentennial Edition of the Works of John Wesley (Nashville: Abingdon Press, 1984), 74.

[23]Moore, "Classical Wesleyanism," 15.

[24]Outler, *Theology in the Wesleyan Spirit*, 23.

[25]L. Tyerman, *The Life and Times of the Rev. John Wesley: Founder of the Methodists*, vol. 2 (London: Hodder and Stoughton, 1870), 401-2.

[26]John Wesley, *The Journal of the Rev. John Wesley*, ed. N. Curnock (Whitefish, MT: Kessinger Publishing, 2006), 60.

[27]Neil Messer, from a Reformed Protestant perspective, gives caution about human sin and

this out very clearly by making a threefold argument:[28] human nature was sinful, which brought about the great flood; human nature has not changed and is even worse since the flood; humans need a remedy for their corrupted nature, and the prescription is holiness. With this argument, he was fighting the Deist view that human nature was positively good. What humanity really needs, according to Wesley, is a "sovereign remedy" to combat our wholly corrupt nature. Still, in light of sociobiological findings over a hundred years later, it seems that Wesley is not too far off from this modern biological perspective. Humans still require some constraints in order to encourage their actions to be altruistic.[29]

Due to his ideas of original sin and justification by faith alone, Wesley was essentially cut off from his Anglican contemporaries;[30] he saw good works and changed behavior as a response to salvation, not a means to salvation. While this understanding of original sin was void of many other elements in traditional Protestant soteriology, it included a catholic doctrine of perfection.[31] As I will discuss later in this chapter, much of Wesley's doctrine of Christian perfection, according to Outler, is an amalgam of various sources—especially, apart from Scripture, the theology of Gregory of Nyssa.[32]

---

sociobiology by noting that "sin might place greater obstacles than we sometimes imagine in the way of our understanding our natural inclinations rightly; another is the possibility that biology (or some other form of empirical investigation) will disclose natural inclinations that are contrary to Christian accounts of the good." Messer, *Selfish Genes and Christian Ethics: Theological and Ethical Reflections on Evolutionary Biology* (London: SCM, 2007), 121.

[28]Wesley, *Sermons 34–70*, 173-85.

[29]Interestingly enough, Neil Messer seems to have sympathetic views to this, despite his Reformed Protestant background. Messer articulately communicates in the conclusion of his *Selfish Gene and Christian Ethics*, "A Christian doctrine of sin can make sense of evolutionary claims and speculations that aspects of our evolutionary inheritance give rise to morally problematic ways of being and acting in the world: the doctrine of sin articulates the insight that God's good creation has become disastrously diverted from its proper ends and goals, so that, as Colin Gunton puts it, only its 'redirection from within' by its creator can restore it to its proper direction." Messer, *Selfish Genes and Christian Ethics*, 247. See also Colin E. Gunton, "The Doctrine of Creation," in *The Cambridge Companion to Christian Doctrine*, ed. Colin E. Gunton (Cambridge: Cambridge University Press, 1997), 143. In this way, as Messer continues, Christians are rightfully able to say at the same time that the world is both "very good" and "by no means as it should be."

[30]Outler, *Theology in the Wesleyan Spirit*, 33.

[31]Ibid.

[32]Ibid., 96n20.

In response to John Taylor's *The Scripture Doctrine of Original Sin: Proposed to Free and Candid Examination,*[33] wherein Taylor expressed, in Wesley's opinion, an overly optimistic view of human nature, Wesley wrote his longest work, which he called, *The Doctrine of Original Sin: According to Scripture, Reason, and Experience.*[34] Here, Wesley combats the new optimism about humanity's innate virtue that denied the Augustinian understanding of the fall and original sin.[35] This issue is significant for Wesley because, as previously mentioned, he views human nature as being totally depraved (yet influenced by prevenient grace). To say otherwise would be to flirt with Pelagianism[36]—a heresy Wesley was certainly against. This long treatise (especially part one) was made more succinct and accessible in his sermon "The Doctrine of Original Sin." For Wesley, there were direct soteriological implications for his followers and he found it necessary to connect the Latin tradition of total depravity (stemming from Augustine) to the Eastern Orthodox understanding of "sin as a disease" and salvation as θεραπεία ψυχῆς (or the treated/healed soul).[37]

As Richard Green notes in *The Works of John and Charles Wesley,* Taylor's work was brought to Wesley's attention years earlier than 1751

---

[33]See John Taylor, *The Scripture-Doctrine of Original Sin: Proposed to Free and Candid Examination in Three Parts* (London: J. Wilson, 1740).

[34]This document was hundreds of pages long and showed the great concern with which Wesley addressed the doctrine of original sin. See John Wesley, *The Doctrine of Original Sin: According to Scripture, Reason, and Experience* (Bristol: F. Farley, 1757). Taylor had his own rebuttal to Wesley, found in his *A Reply to the Reverend Mr. John Wesley's Remarks on the Scripture-Doctrine of Original Sin to Which Is Added, a Short Inquiry into the Scripture-Sense of the Word Grace* (London: M. Waugh, 1767).

[35]Wesley, *Sermons 34–70,* 170.

[36]Albert Outler makes a keen observation about the dangers of Pelagianism and the threat of heresy that can occur when one is distancing oneself from Pelagianism. In his *Theology in the Wesleyan Spirit,* Outler notes that Wesley found himself in a precarious position, "For if you argue that we are sinful by nature (i.e., that the power only to sin is the actual human condition), you are also on the verge of saying that the original sin is simply being human—and that's heresy. If you take the opposite side, and argue that we *can* banish sin from our own lives and societies whenever we muster up sufficient moral effort (prodded inwardly by conscience and outwardly by moral example and admonition), you are on the verge of saying that sin is, in essence, a sort of social dysfunction, corrigible by moral insight and effort, or by proper programs of social reform. If you then persist in arguing for *original* sin, in some sense or other, you may be implying that we are sort of badly botched animals since, clearly, no other animal 'sins' with anything like the same regularity, recklessness, and tragic consequence as does the human animal. But this is heresy as well—for it denies the moral uniqueness of the human creation." See Outler, *Theology in the Wesleyan Spirit,* 25.

[37]Wesley, *Sermons 34–70,* 171.

and Wesley carefully considered it, hoping that "someone of position and competent ability would answer it"—but Wesley grew impatient.[38] Instead of waiting, he records in his journal, dated Wednesday, April 10, 1751, "I rode to Shackerley. Being now in the very midst of Mr. Taylor's disciples, I enlarged much more than I am accustomed to do, on the doctrine of Original Sin; and determined, if God should give me a few years' life, publicly to answer his new gospel."[39] According to Thomas Oden,

> Wesley thought that Taylor was working out of a deistic theism, a Pelagian anthropology, a reductionist Christology, a work-righteousness ethic, and a universalist eschatology, all of which were undermining substantive Christian teaching. Wesley considered Taylor's unitarianism as tending toward antinomianism, toward the trivializing of Christ's work on the cross, the weakening of Christ's deity, and finally the impugning of God's character by making God responsible for present human sinfulness.[40]

In his 1740 work, Taylor posits, concerning sinfulness, that humans can choose to do otherwise: "They can do their duty *if* they choose!"[41] Wesley strongly disagreed with this notion of seemingly unadulterated will, arguing instead that there was no good in human nature alone, apart from the prevenient grace of God.[42] Wesley stated, "I verily believe no single person since Mahomet has given such a wound to Christianity as Dr. Taylor. . . . [He has] poisoned so many of the clergy, and indeed the

---

[38]Richard Green, *The Works of John and Charles Wesley: A Bibliography Containing an Exact Account of All the Publications Issued by the Brothers Wesley, Arranged in Chronological Order, with a List of the Early Editions, and Descriptive and Illustrative Notes*, 2nd ed. (London: Methodist Publishing House, 1906).

[39]John Wesley, 2:226.

[40]Thomas C. Oden, *John Wesley's Scriptural Christianity: A Plain Exposition of His Teaching on Christian Doctrine* (Grand Rapids: Zondervan, 1994), 159.

[41]Taylor, *The Scripture-Doctrine of Original Sin.* Emphasis his. Yet, it should be noted that Taylor believed it is not a mere "even choice" that one should just "choose."

[42]In sermon 44, "Original Sin," Wesley states, "But was there not good mingled with the evil was there not light intermixed with the darkness No; none at all: 'God saw that the whole imagination of the heart of man was only evil.' It cannot indeed be denied, but many of them, perhaps all, had good motions put into their hearts; for the Spirit of God did then also 'strive with man,' if haply he might repent, more especially during that gracious reprieve, the hundred and twenty years, while the ark was preparing. But still "in his flesh dwelt no good thing;" all his nature was purely evil: It was wholly consistent with itself, and unmixed with anything of an opposite nature." Wesley, *Sermons 34–70,* 175.

fountains themselves—universities in England, Scotland, Holland, and Germany."[43]

In response to Pelagian optimism and Augustinian pessimism concerning the human flaw and the human potential, Wesley desired an alternative.[44] It was unlike him to commit to an extreme position, including the Augustinian-Calvinist position of total depravity—rather he chose a "not yet" *via media*,[45] combining the idea of total depravity with the idea of prevenient grace. Much of Wesley's solution comes from his sermon "Original Sin": God's method of healing a soul that is diseased is his θεραπεία ψυχῆς (God's healing activity within our hearts); God is universal in calling sinners to be healed from sin (which functions as a sickness within the human person). This idea on original sin, then, becomes a counterweight to "election." Yet, one has to accept God's prescription, which includes repentance, renunciation of self-will and trust in God's unmerited grace. To this end, Wesley concludes at the end of "Original Sin," "Ye know that the great end of religion is, to renew our hearts in the image of God, to repair that total loss of righteousness and the true holiness which we sustained by the sin of our first parent."[46] Thus, all humans are born with a totally depraved human nature—it just so happens that no human is without the prevenient grace of God.[47]

---

[43]John Wesley, *The Letters of the Rev. John Wesley*, ed. John Telford, standard ed., 8 vols. (London: Epworth Press, 1931), 4:48.

[44]Outler, *Theology in the Wesleyan Spirit*, 35. One limitation for Wesley is that he accepts the framework of Augustine on original sin *and* the Eastern framework on the *imago Dei* and perfection. This becomes a tension for Wesley that I will address in my conclusion.

[45]Outler expresses a cogent framework for how prevenient grace fits in to the doctrine of total depravity. I draw from that in the following paragraph. See ibid., 37.

[46]Wesley, *Sermons 34–70*, 185.

[47]Outler articulates well the unique contribution that Wesley brings to this doctrine. Wesley still upholds the sovereignty of grace but notes that it is *not* irresistible. "Sinners," says Outler, "can do literally nothing to save themselves (not by merit, nor demerit, nor by the will to believe). And yet God's intention in creating persons (which gives each person his/her unique identity) is not thwarted by human resistance, because it is God's own purpose that the offer of grace shall be experienced as *optional*. The chief function of prevenient grace, therefore, is to stir the sinner to repentance (which is to say, to a valid self-understanding of his/her sinfulness). Thus, Wesley can speak of repentance as the *porch* of religion, of faith as the *door*, and of holiness as religion itself." See Outler, *Theology in the Wesleyan Spirit*, 37-38. Consequently, one can see how both God and humans take part in this process. Humans do the repenting, and God gives unmerited grace. Also, Wesley clearly thought that Christians are capable of not sinning; to hold otherwise is Old Testament thinking under the law. Christians live under a new dispensation where Scripture confirms the "necessity of sinning no longer exists." See Moore, "Development in Wesley's

It is this grace that allows Wesley to be pessimistic toward human nature, but optimistic about the human condition. As Outler points out in *Theology in the Wesleyan Spirit*, it is here that Wesley turns away from the common Protestant doctrines of salvation. The Reformed tradition (Luther and Calvin in particular) "regarded the residue of sin (*fomes peccati*) not only as ineradicable but sinful as such; it falls under God's righteous condemnation even though this does not forfeit his justifying grace."[48] Yet Wesley differentiated between "sin properly so-called" (a deliberate and conscious violation of a known law of God) and involuntary transgressions that "[presuppose] that residual sin (*fomes peccati*) diminishes in force and influence as the Christian grows in grace."[49]

5.1.2 *Wesley's understanding of perfection*

To understand John Wesley's motivation in regard to crafting the theological concept of Christian perfection, it is important to take note of his intellectual influences during the time this concept was developed. Not only was Wesley dramatically influenced by Jeremy Taylor's *Rules and Exercises of Holy Living* and *Rules and Exercises of Holy Dying*,[50] but also by Thomas à Kempis's *The Imitation of Christ*.[51] Both Taylor's and Kempis's books focused on Christian perfection—Taylor's emphasizing perfection *in this life* and Kempis's focusing on perfection in the *life to come*.[52] Wesley ultimately rejected Kempis's understanding of perfection because he believed that God made humans capable of experiencing happiness and good things on this earth, such as perfection.[53] This critical distinction of Taylor's "in this life" understanding of perfection (over Kempis's "life to come" understanding), pointing toward the attainability of Christian perfection in this life, launched Wesley on a path that caused great controversy not only in his own time, but also in ours.

Thought on Sanctification and Perfection," 36. Wesley notes, "A Christian is so far perfect as not to commit sin." See Wesley, *John Wesley: A Representative Collection of His Writings*, 267.

[48]Outler, *Theology in the Wesleyan Spirit*, 38.

[49]Ibid.

[50]Jeremy Taylor, *Selected Works*, ed. Thomas K. Carroll, Classics of Western Spirituality (New York: Paulist Press, 1990), 439-504.

[51]Thomas à Kempis, *The Imitation of Christ*, ed. Aloysius Croft and Harold Bolton, Dover Thrift Editions (Mineola, NY: Dover Publications, 2003).

[52]Moore, "Development in Wesley's Thought on Sanctification and Perfection," 30.

[53]For more on this, see Martin Schmidt, *John Wesley: A Theological Biography* (Nashville: Abingdon Press, 1963), 77-80.

Wesley started to shape his understanding of perfection (or entire sanctification) after his adoption of Taylor's interpretation of perfection. To put it simply, if the concepts of justification and regeneration[54] explain what God does *for* us, then sanctification and entire sanctification define what God does *in* us. According to Outler, entire sanctification causes one to "mature and fulfill the human potential according to his primal design."[55] Yet, as Outler continues, "few Christians had ever denied some such prospect, *in statu gloriae*; few, in the West at least, had ever envisioned it as a realistic possibility *in this life*."[56] This emphasis is what made the concept of Christian perfection difficult for Wesley's contemporaries because it seemed to go against the "grace based" theology of both justification and regeneration. Still, as Outler also points out, "'perfect love,' as Wesley understood it, is the conscious certainty, *in a present moment*, of the fullness of one's love for God and neighbor, as this love has been initiated and fulfilled by God's gifts of faith, hope and love. This is not a state but a dynamic process: saving faith is its beginning; sanctification is its proper climax."[57] This very nuanced understanding of such perfection betrays the concept through which Wesley sees perfection: the Eastern understanding of θέωσις (hereafter *theosis*).[58]

Wesley uses the phrase *image of God* to describe the human ability for knowing and the human capability to respond to God's grace (namely

---

[54]In the Wesleyan tradition, *justification* typically denotes the act of one attaining salvation through faith in the grace of Jesus Christ. *Regeneration*, in the Wesleyan tradition, typically signifies the beginning of a process by which individuals progress in Christian faith.

[55]Wesley, *Sermons 34–70*, 97. Emphasis his.

[56]Ibid. Emphasis his.

[57]Wesley, *John Wesley: A Representative Collection of His Writings*, 31. Emphasis his.

[58]It should be stated that this restoration of the *imago Dei* within Christian perfection might come about as an "instantaneous" change or through a process resulting in a momentary change that is difficult to pinpoint. In *A Plain Account of Christian Perfection*, Wesley states, "An instantaneous change has been wrought in some believers: None can deny this. Since that change they enjoy perfect love; they feel this, and this alone; they 'rejoice evermore, pray without ceasing, and in everything give thanks.' Now, this is all that I mean by perfection; therefore, these are witnesses of the perfection which I preach. But in some this change was not instantaneous. They did not perceive the instant when it was wrought. It is often difficult to perceive the instant when a man dies; yet there is an instant in which life ceases. And if ever sin ceases, there must be a last moment of its existence, and a first moment of our deliverance from it." See Wesley, *The Works of John Wesley*, 11:442.

God's prevenient, justifying and sanctifying grace).[59] Under *theosis*, humanity's corrupted and disabled image is restored to its fullness; thus, *theosis* becomes the goal of Wesley's *ordo salutis*.[60] The recovery of this image of God for Wesley was significant because, as Theodore Runyon observes, "the renewal of the image functions in a way similar to the Eastern Fathers' doctrine of *theosis* which, whether it describes the beginning of the journey of faith or its culmination, is effective participation in divine reality which both guides the believer at every step along the way and culminates the journey."[61] However, tying his theology to Orthodoxy did cause backlash from Wesley's contemporaries. Those influenced by the idea of the *simul justus et peccator* (simultaneously both righteous and a sinner)—where translations of *perfectio* were viewed as a "perfected perfection"—understood the concept of *theosis* in Christian perfection as "works righteousness" and anathema to proper Protestant doctrine.[62] In Outler's preface to Wesley's sermon on Christian perfection, he states,

> Even the Methodists, working from their own unexamined Latin traditions
> of forensic righteousness, tended to interpret "perfection" in terms of a spir-
> itual elitism—and so misunderstood Wesley and the early Eastern traditions
> of τελειότης (perfection or perfectness) as a never ending aspiration for all of
> love's fullness (perfecting perfection). Thus, "Christian Perfection" came to
> be the most distinctive and also the most widely misunderstood of all Wes-
> ley's doctrines.[63]

This was so problematic for Wesley that Bishop Edmund Gibson found it necessary to look into Wesley's teaching on perfection (at the time, the Methodists were housed within his diocese).[64] Wesley was very open with the bishop and recounted the event in *A Plain Account of Christian Perfection*, stating, "I told him, without any disguise or reserve. When I ceased speaking, he said, 'Mr. Wesley, if this be all you mean, publish it

---

[59]Wesley, *Sermons 1–33*, 117-18n5.
[60]Ibid.
[61]Theodore Runyon, "The New Creation: The Wesleyan Distinctive," *Wesley Theological Journal*
    31, no. 2 (1996): 14.
[62]Wesley, *Sermons 34–70*, 98.
[63]Ibid.
[64]Ibid., 97.

to all the world. If anyone then can confute what you say, he may have free leave.' I answered, 'My Lord, I will' and accordingly wrote and published the sermon on Christian perfection."[65] Despite having formal approval from a bishop and giving relentless explanations, Wesley found himself defending his understanding of Christian perfection for most of his life.

The concept of *theosis*, literally meaning "ingoddedness" or "becoming god," is found in the Eastern Orthodox tradition,[66] with which Wesley was quite familiar. As Michael Christensen mentions in "Theosis and Sanctification: John Wesley's Reformulation of a Patristic Doctrine," the notion of *theosis* embodies a "vision of human potential for perfection, anticipated in ancient Greece, witnessed to in both the Old and New Testaments, and developed by Patristic Christian theologians of the first five centuries after Christ . . . and persists yet today in Eastern Christianity as a challenge to Western theology."[67] To be sure, *theosis* is best understood as gift from God, much like justifying and regenerative faith.[68]

Despite being persistent to his critics that Christian perfection was not "works righteousness" but rather a form of grace, Wesley struggled to convince followers to "work on perfection" while accepting the paradoxical tension between grace and works. In his introduction to Wesley's sermon 40, "Christian Perfection," Outler states, "Wesley's encouragement to his people to 'go on to perfection' and to '*expect* to be made perfect in love in this life' aroused lively fears that this would foster more of the self-righteous perfection*ism* already made objectionable by earlier pietists."[69] So prominent was this struggle that Wesley warns against one who takes on a standard of perfection higher than what Wesley calls "Scripture perfection." In "A Plain Account of Christian Perfection," he states, "Scripture perfection is: pure love filling the heart, and governing all the words and actions. If your idea includes anything more or anything else, it is not scriptural; and then no wonder, that a scripturally perfect

---

[65] Wesley, *The Works of John Wesley*, 11:374.
[66] Christensen, "Theosis and Sanctification," 72.
[67] Ibid.
[68] Wesley, *John Wesley: A Representative Collection of His Writings*, 253.
[69] Wesley, *Sermons 34–70*, 97. Emphasis Outler's.

Christian does not come up to it."[70] However a high standard this seems to be, this is the measure by which Wesley communicates his doctrine: a kind of holiness buttressed by the Eastern concept of *theosis*, which emphasizes both grace and works. Yet Wesley makes clear that there is a role for the Holy Spirit. In sermon 39, "Catholic Spirit," Wesley writes, "Is thy faith ἐνεργουμένη δι' ἀγάπης—filled with the energy of love?"[71] Runyon's summary is fitting as he comments, "On the basis of human efforts alone, this kind of self-giving love is impossible. But the source of the energy is the love of God received through the life-giving Spirit."[72]

Perhaps one of the challenges with Wesley's views of *theosis* is that he is frequently misinterpreted by people who understand the word *perfect* to be translated literally from the Latin *perfectio* (where in Medieval Latin *perfectus* meant "faultless" or "unable to be improved").[73] Yet, despite these misunderstandings of what Wesley was actually claiming, he remained steadfast in his presentation. As Outler puts it, "[Wesley] was hard pressed to explain [perfection] to both disciples and critics; he sought earnestly to correct its misinterpretations by the cynics, on the one side, and the fanatics on the other. But he seems never to have felt seriously moved either to abandon the doctrine or to modify it to suit his objectors."[74] Wesley even implored numerous examples and anecdotes to persuade his critics about this difficult concept. In "A Plain Account of Christian Perfection" Wesley raises an interesting hypothetical situation:

> Q. But if two perfect Christians had children, how could they be born in sin, since there was none in the parents? A. It is a possible, but not a probable, case; I doubt whether it ever was or ever will be. But waving this, I answer: Sin is entailed upon me, not by immediate generation, but by my first parent. "In Adam all died; by the disobedience of one, all men were made sinners." All men, without exception, who were in his loins when he ate the forbidden fruit.[75]

[70]Wesley, *The Works of John Wesley*, 11:401.
[71]Wesley, *Sermons 34–70*, 88.
[72]Runyon, "The New Creation," 15.
[73]Wesley, *John Wesley: A Representative Collection of His Writings*, 30.
[74]Ibid., 253.
[75]Wesley, *The Works of John Wesley*, 11:400.

Thus, despite Wesley's attempt to communicate the idea of Christian perfection through numerous methods, his detractors remained skeptical of his claims, which led to a longstanding battle for him.

Yet still, this kind of Christian perfection, embattled as it was, continued to be invoked by Wesley with a particular nuance: perfection is to be attained *in this life*; nowhere do we find in Wesley's writings that Christian perfection is merely for the life to come. As Outler expounds, "If Wesley's writings on perfection are to be read with understanding, his affirmative notion of 'holiness' *in the world* must be taken seriously—active holiness in *this* life—and it becomes intelligible only in the light of its indirect sources in early and Eastern spirituality."[76] Recognizing how steadfast Wesley was in urging perfection in this life is important for understanding the tone he set in his bands and classes (as discussed in chapter six). Despite the tension in language often associated with ideas of perfection, Wesley continued to teach Christian perfection (a concept full of implications for personal ethics and social transformation), always staying consistent with his original position.[77] Given the landscape of that with which Wesley was dealing, he would have had fewer critics if the idea of Christian perfection had been trumpeted "as the Christian ideal to be realized *in statu gloriae*—or if the doctrine had followed the classical Protestant line that justification and sanctification are two aspects of the same thing: God's pardoning grace."[78] But convenience was not on Wesley's itinerary; instead, he chose to be truthful to his understanding of biblical Christianity, particularly to his understanding of Matthew 5:48. So truthful was Wesley that he counted the cost of his words, cautioning others who would mention the concept of Christian perfection. In this way, Wesley warns followers about the radical view of preaching perfection in this life. As he states in the first

---

[76]See Wesley, *John Wesley: A Representative Collection of His Writings*, 252. Wesley found inspiration for the potentiality of perfection in Origen's *Dialogue with Heraclides*: "I beseech you, therefore, be transformed. Resolve to know that in you there is a capacity to be transformed." See Origen, "Dialogue with Heraclides," in *Alexandrian Christianity*, ed. John Ernest Leonard Oulton and Henry Chadwick, The Library of Christian Classics (Philadelphia: Westminster Press, 1954), 150. It should also be stated, as Michael Christensen points out, that Wesley is cautious to give assent to all of Origen's works. See Christensen, "Theosis and Sanctification," 79-80.

[77]Wesley, *Sermons 34-70*, 98-99.

[78]Wesley, *John Wesley: A Representative Collection of His Writings*, 253.

line of sermon 40, "Christian Perfection," "There is scarce any expression in Holy Writ which has given more offence than this. The word 'perfect' is what many cannot bear. The very sound of it is an abomination to them. And whosoever 'preaches perfection' (as that phrase is), i.e. asserts that it is attainable in this life, runs great hazard of being accounted by them worse than a heathen man or a publican."[79] Yet despite all the controversy, he remained faithful to his cause and used such flaming rhetoric.

For Wesley, part of being truthful to the biblical understanding of perfection was recognizing the connection between Christian perfection and what he called "social holiness," where inward holiness resulted in outward action. When talking about Christian perfection and social holiness, Wesley notes, "The gospel of Christ knows of no religion, but social; no holiness but social holiness. 'Faith working by love' is the length and breadth and depth and height of Christian perfection."[80] Herein was his point of origin for the topic of Christian perfection. As Runyon concludes, "He never tires of reminding us that perfection is nothing greater and nothing less than 'loving God with all our heart, and our neighbor as ourselves.'"[81] Furthermore, as Outler states in his introductory commentary for sermon 40, "Christian Perfection," "If, for Wesley, salvation was the total restoration of the deformed image of God in us, and if its fullness was the recovery of our negative power not to sin and our positive power to love God supremely, this denotes that furthest reach of grace and its triumphs in this life that Wesley chose to call 'Christian Perfection.'"[82] Wesley says in sermon 17, "The Circumcision of the Heart":

> I am, first, to inquire, wherein that circumcision of the heart consists, which
> will receive the praise of God. In general we may observe, it is that habitual
> disposition of soul which, in the sacred writings, is termed holiness; and
> which directly implies, the being cleansed from sin, "from all filthiness both
> of flesh and spirit;" and, by consequence, the being endued with those virtues

---

[79]Wesley, *Sermons 34–70*, 99.
[80]John Wesley, *The Works of John Wesley*, 14:321.
[81]Runyon, "The New Creation," 15.
[82]Wesley, *Sermons 34–70*, 97.

which were also in Christ Jesus; the being so "renewed in the spirit of our mind," as to be "perfect as our Father in heaven is perfect."[83]

Wesley also concludes with a list that portrays what a "circumcised heart" looks like. Here, his remarks on perfection sound very much like Taylor and Law's understanding of the concept.[84] Yet still, Wesley habitually notes that if Christian perfection were possible to attain, it would *have to be* possible to attain *within this life* where social holiness would be its fruit. This kind of perfection would remain consistent with not only the whole of a person's spiritual experience, but also Wesleyan soteriology in general.[85] Perfection in the Wesleyan context, then, is a kind of "voluntary death" where the individual dies to all self-interest and only seeks to please God.[86] Thus, as Marselle Moore notes, "Instead of stating

---

[83]The following is the above-mentioned list taken from "A Plain Account of Christian Perfection" as well as sermon 17, "The Circumcision of the Heart." See Wesley, *The Works of John Wesley*, 11:367-68. See also Wesley, *Sermons 1–33*, 401-14. Wesley states, "Here, then, is the sum of the perfect law; this is the true circumcision of the heart. Let the spirit return to God that gave it, with the whole train of its affections. 'Unto the place from whence all the rivers came, thither' let them flow again. Other sacrifices from us he would not; but the living sacrifice of the heart he hath chosen. Let it be continual offered up to God through Christ, in flames of holy love. And let no creature be suffered to share with him: For he is a jealous God. His throne will he not divide with another: He will reign without a rival. Be no design, no desire admitted there, but what has Him for its ultimate object. This is the way where in those children of God once walked, who, being dead, still speak to us: 'Desire not to live, but to praise his name: Let all your thoughts, words, and works, tend to his glory. Set your heart firm on him, and on other things only as they are in and from him. Let your soul be filled with so entire a love of him, that you may love nothing but for his sake.' 'Have a pure intention of heart, a steadfast regard to his glory in all your actions.' 'Fix your eye upon the blessed hope of your calling, and make all the things of the world minister unto it.' For then, and not till then is that 'mind in us which was also in Christ Jesus,' when, in every motion of our heart, in every word of our tongue, in every work of our hands, we 'pursue nothing but in relation to him, and in subordination to his pleasure'; when we, too, neither think, nor speak, nor act, to fulfil our 'own will, but the will of him that sent us'; when, whether we 'eat, or drink, or whatever we do, we do all to the glory of God.'"

[84]Moore, "Development in Wesley's Thought on Sanctification and Perfection," 32.

[85]Wesley, *John Wesley: A Representative Collection of His Writings*, 253. In *Principles of a Methodist*, Wesley also mentions that the "general prejudice against Christian perfection may chiefly arise from a misapprehension of the nature of it. We willingly allow, and continually declare, there is no such perfection in this life, as implies either a dispensation from doing good and attending all the ordinances of God; or a freedom from ignorance, mistake, temptation, and a thousand infirmities necessarily connected with flesh and blood." See John Wesley, *The Works of John Wesley*, 8:364. For a more comprehensive look at how Wesley used the idea of perfection in *Principles of a Methodist*, see appendix one.

[86]Moore, "Development in Wesley's Thought on Sanctification and Perfection," 31. Moore also mentions that this kind of perfection Wesley mentions is not freedom from making mistakes. See ibid., 35.

what perfection is, Wesley tells us who is perfect. Using scriptural phrases which become stock phrases, he says one is perfect who has 'the mind which was in Christ,' who 'walketh as Christ walked,' who is 'cleansed from all filthiness of flesh and spirit,' who 'doth not commit sin,' etc."[87] So focused on the idea that inward holiness leading to Christian perfection was just as much an act of grace as justification or regeneration, Wesley viewed the doctrine of perfection as "yet another way of celebrating the sovereignty of grace."[88]

5.1.3 A spiritual component

In Wesleyan theology, any moment of progression for a follower has both a natural component (humans working toward God) and a spiritual component (grace being imparted by God to humans). It is perhaps no coincidence that Wesley's own life reflects such components. It has long been held that John Wesley's full conversion came during his experience at an Aldersgate Street meeting, where Peter Böhler had formed several bands.[89] Wesley was listening to Luther's "Preface to the Book of Romans" being read and he recalls in his journal on May 24, 1738, "About a quarter before nine, while [Luther] was describing the change which God works in the heart through faith in Christ, I felt my heart strangely warmed. I felt I did trust in Christ, Christ alone for salvation, and an assurance was given me that he had taken away *my* sins, even *mine,* and saved *me* from the law of sin and death."[90] This would be a defining moment in Wesley's life,[91] when personal experience with God moved him to live an inwardly changed life that resulted in outward fruit. Yet, though Wesley did, indeed, experience an inwardly changed life, the outward fruit of such would not be evident until he encountered even more radical religious experiences—experiences that helped to solidify the movement, providing it with an experiential foundation leading toward both inward and outward holiness. If individuals are totally depraved (apart from the

---

[87]Ibid., 34-35. See also Wesley, *The Works of John Wesley*, 8:374.

[88]Wesley, *John Wesley: A Representative Collection of His Writings*, 253.

[89]John Wesley, *Journal and Diaries*, ed. Richard P. Heitzenrater and W. Reginald Ward, The Bicentennial Edition of the Works of John Wesley (Nashville: Abingdon Press, 1988), 249n75.

[90]Ibid., 249-50.

[91]Kenneth J. Collins does a nice job of illustrating this importance. See Kenneth J. Collins, *John Wesley: A Theological Journey* (Nashville: Abingdon Press, 2003), 86-90.

prevenient grace of God), then they need God working in them through their experiences to be able to overcome biological constraints (or, as Wesley would call it, "human nature"). Wesley saw this as a key and critical component in his development of the early Methodists, and I will recall a few of these occurrences.

If Aldersgate is the spiritual birth of the movement, the temporal birth of the revival took place in a meeting on Monday, January 1, 1739—a meeting that became the catalyst for revival; for it is out of this meeting that George Whitefield began converting so many individuals that Wesley was called to come help. On that Monday, Wesley recounted in his journal,

> Mr. Hall, Kitchin, Ingham, Whitefield, Hutchins, and my brother Charles were present at our love-feast in Fetter Lane, with about sixty of our brethren. About three in the morning, as we were continuing instant in prayer, the power of God came mightily upon us, insomuch that many cried out for exceeding joy, and many fell to the ground. As soon as we were recovered a little from that awe and amazement at the presence of his Majesty, we broke out with one voice, "We praise thee, O God; we acknowledge thee to be the Lord."[92]

As one can see, Wesley and Whitefield had a religious experience that was so potent it demanded "recovery" time.

Nearly a year after his Aldersgate conversion, Wesley first took part in the new revival movement. In a journal entry from April 1739, Wesley mentioned that he was "sensibly led" to go help the revival. His rhetoric was both logical and spiritual (with words like *thunderstruck*). On Thursday, April 26, 1739, Wesley wrote, "I was sensibly led, without any previous design, to declare strongly and explicitly [a message from God]. Immediately one and another and another sunk to the earth: they dropped on every side as thunderstruck. One of them cried aloud. We besought God in her behalf, and he turned her heaviness into joy. A second being in the same agony, we called upon God for her also, and he spoke peace unto her soul."[93]

---

[92]John Wesley, *Journal and Diaries*, 29.

[93]Ibid., 51. Just before this instance on Saturday, April 21, 1739, a similar occurrence happened. Wesley writes, "At Weavers' Hall a young man was suddenly seized with violent trembling all over, and in a few minutes, 'the sorrows of his heart being enlarged,' sunk down to the ground.

One reason these stories are so important—and the first hundred pages of W. R. Ward's and Richard Heitzenrater's edited volume, *Journals and Diaries*,[94] are swollen with anecdotes—is because they express Wesley's theological idea that God *has to be* involved in the process of changing individuals. This goes back to Wesley's understanding of original sin. For Wesley, God was at work in the early Methodist movement. Therefore, none of the individuals in his bands and classes were relegated to total depravity apart from prevenient grace. In his own understanding, this meant Wesley's followers were not solely bent toward egoism—allowing them to be molded by a structure of constraints toward a kind of inward holiness that led to outward action (in specifically, altruistic action). This idea is critical.

5.2 GENETIC SELFISHNESS AND ITS IMPLICATIONS FOR WESLEYAN ETHICS

The idea of genetic selfishness that became prevalent in the mid- to late twentieth century should make Wesleyan ethicists pause. In fact, it is difficult to give assent to evolutionary theory and have knowledge of the Wesleyan concept of Christian perfection without experiencing some dissonance rooted in the seemingly contradictory points of contention: Christian perfection and humanity's biological constraints. We would do well to note that this apparent dilemma is much like Wesley's attempt to wed a theology of original sin with a theology of Christian perfection. Nevertheless, as observed earlier, Wesley reconciled such differences by means of the notion of prevenient grace.[95] Similarly, it is my contention

---

But we ceased not calling upon God, till he raised him up, full of 'peace and joy in the holy Ghost.'" See ibid., 50.

[94] Ibid.

[95] There has been much debate over whether Wesley's concept of prevenient grace should be considered semipelagianism (a concept that avoids the heresy of Pelagianism without having to adopt a strict Augustinian approach to free will and grace). However, such a distinction is beyond the scope of this book; therefore, I will instead be using the concept of prevenient grace as Wesley formed it. For more on this discussion, see Louis Berkhof, *Systematic Theology* (Grand Rapids: Eerdmans, 1996), 219-26. It should also be noted that, concerning this concept of prevenient grace, Ken Collins articulates five attributes that benefit individuals in his *Theology of John Wesley*: the individual gains basic knowledge of the attributes of God; the individual gains a reinscription of the moral law; the individual gains a conscience; the individual gains a measure of free will that is graciously restored; wickedness is restrained. See Kenneth J. Collins, *The Theology of John Wesley: Holy Love and the Shape of Grace* (Nashville: Abingdon Press, 2007), 78.

that grace is the solution to the dilemma of Christian perfection and constraints from human evolution (including selfish and selfless tendencies). After all, Christian perfection in Wesleyan thought, much like the restoration of original sin, is a means of grace just as justification or regeneration grace is.

To this end, there are three primary propositions we might make when attempting to provide a solution for how one may attain perfection *in this life* while still being bound by the same genetic code with which one is born. First, one might argue that the biological makeup of a person is *replaced* with a different genetic composition after the individual attains perfection. Such a proposition seems highly improbable since many scholars would see such a replacement as contrary to God's character. Due to the fact that biological traits are permanent, at least to the extent that science can prove such, many Wesley scholars, such as Howard Snyder, assert that God works *within* creation and even creation's limits.[96] It would follow, then, that it is highly unlikely for God to simply replace genetics.

In fact, a total "replacement approach" raises important questions about whether the person who is redeemed is the same person as the one who was fallen. This problem leads to a cyclical conundrum. For instance, before justifying faith an individual is heavily influenced by her or his genetic inclinations. Working toward Christ and receiving grace from God can justify this individual. Yet, if the individual becomes a totally different person subsequent to her or his justification, either genetically or even "in essence," it would follow that the "justified person" is not the same individual as the one who was in need of justification. Do they now need to be justified again? Hence we can see a cyclical problem.

Another major problem with this replacement approach is that this splitting of the soul and body into distinct aspects of the person flirts with dualism. It bears certain relations to gnosticism, which denies the goodness of the body. In this heretical view, the gnostic desires to escape

---

[96]See Howard A. Snyder and Joel Scandrett, *Salvation Means Creation Healed: The Ecology of Sin and Grace; Overcoming the Divorce Between Earth and Heaven* (Eugene, OR: Cascade Books, 2011).

from the captivity of the body, and genetic material is not something made wholly good by the Creator. In this way, the replacement approach leads to a slippery slope by which the fundamental doctrines of the church—such as the resurrection, dual nature of Christ and the virgin birth—are in jeopardy. If the body is merely something to *retreat from* and be ultimately *replaced by* a "perfected material essence," then we are left with the incompatible ideas of a Creator God (who is active in the world as evidenced in the Old and New Testaments) who is materially absent from this new creation. Still another difficulty with the replacement approach rests in the following question: How can an individual whose denigrated genetic material, which has been replaced by some kind of suprascientific essence, ever fall from grace? If one is in a state of perfection that is not corrupted or even influenced by genetic material, it does not seem possible for the individual to change back to its previous state. Instead, it would be assumed that the replaced material would not be an influencing factor, but rather permanent; thus, once an individual attained perfection it would be everlasting.

The second proposition working toward the remedy of the sociobiology problem is directly tied to a dualistic/gnostic interpretation: it is possible for the body to remain in a corrupt state of selfishness while the soul achieves Christian perfection in this life. If this proposition were correct, it would also have major implications for the orthodox theologies of the virgin birth, incarnation and the bodily resurrection of Christ. One can assume that if the material body is still wholly intact, with genetic proclivities influencing an individual toward selfish behavior, then this material body would still have the same influence on the soul. For example, an individual with years of drug addiction whose soul is perfected must remain in the addicted body. It would make sense, then, that this material body, with its chemical and genetic influences, would necessarily corrupt the state of the soul. There seems to be few valid arguments where the soul could be uninfluenced by the body if they remained parallel.[97] Contemporary materialists, such as Paul and Patricia Churchland, undoubtedly reject ideas of a parallel working of

---

[97]This is due to the undoubted continuous interaction between the two.

the body and soul, citing clear connections between the body and what some consider the soul.[98] For materialists, and many within the scientific community, the concept of the "soul" is itself called into question, and to say that the body and soul work in parallel would be a stretch. Moreover, ideas of parallelism would go against many in the theological community as well, showing clear ways God has decided to work *within* creation rather than parallel with it.[99] For these reasons, the second proposition, the body and perfected Christian working in parallel, is also highly unlikely.[100]

It is precisely because of the inadequacy of the first two explanations that another solution should be posited that does not reside in the dualism of the body and soul, yet satisfies the puzzle of Christian perfection and humanity's biological makeup. It is thus in the third proposition where we may find a solution to our problem: through a combination of human choice and the grace of God, humans possess the ability to continually "overcome"[101] their genes and achieve Christian perfection in

---

[98]To be fair, in the first few pages of *Braintrust*, Patricia Churchland sets up a straw man anecdote about the absurdity of God and morality—discussing medieval witch trials and the like. However, she makes very good points of the connectivity between the material body and what some consider the "soul," showing that we are, in fact, one entity. See Patricia S. Churchland, *Braintrust: What Neuroscience Tells Us About Morality* (Princeton, NJ: Princeton University Press, 2011), 1-2.

[99]See again Snyder and Scandrett, *Salvation Means Creation Healed*. There are also numerous references from Scripture that expose God working *within* creation. One can easily find selections of hundreds of passages with a quick search. These passages range from the Genesis accounts of creation, to God using a flood as a form of justice, to the incarnation and bodily resurrection of Jesus. Most denominations in the Wesleyan tradition, in fact, have language in their books of discipline concerning an active God who is not only the "creator" but also the "sustainer" of the universe.

[100]One might also be able to argue that the opposite of parallelism would be an absent Deist God who does not work within creation at all. Yet this notion would run contrary to much Christian Scripture and Christian tradition. See previous footnote as an example.

[101]When using the word *overcome* I always mean to confer the sense of the "continual process of overcoming." By doing so, I hope to draw attention to both the present progressive *and* future progressive nuances of the verb *overcoming*. At the same time, I do not want to mitigate the present perfect nature of the word *overcome*, which helps us understand the grace that helps the individual overcome biological proclivities toward violence, for instance, both now and progressively in the future. Thus, owing to its perspectival nuances, I have chosen to employ the word to describe how a Christian can become perfect *in this life* without being determined by her or his biology; for I believe this process embodies the heart of Christian perfection. A perfected Christian is not uninfluenced by biology, but, through the working of grace, is able to overcome the limits of biological influences. I have chosen *overcome* instead of *transcend* because I do not wish to convey the concept of "otherworldliness" or to fall into gnosticism.

this life. Such grace is much the same as prevenient grace as it works alongside the theological concepts of original sin, justifying and regenerative grace, aiding the believer during and after the free conversion, and sanctifying grace working with Christians as they move toward perfection. In the same way, grace helps a believer achieve Christian perfection in this life. With the use of *overcome*, I mean the connotation that it is a grace that helps one go beyond the limitations and propensities of one's genes, a grace that allows one to continually overcome her or his constant proclivity toward selfishness. Through this process, human actions and choices are not determined by innate tendencies, passions or biological predispositions.[102] Yet the genetic urges toward selfishness do not dissipate or go away; one merely has the grace to overcome the genetic urges and achieve Christian perfection, as well as the grace to continue in the process of overcoming the genetic urges after perfection.

In Wesleyan theology, an individual's works are always a *response* to grace. One is never able to work hard enough to attain either justification or entire sanctification. Instead, just as prevenient grace beckons individuals toward a justifying Savior, so does this sanctifying grace call the Christian to respond to the Holy Spirit's urgings. At some point, the individual becomes more influenced by the Holy Spirit's persuasions than genetic proclivities, moving them further down the path of sanctification and ultimately enabling the individual to overcome those genetic constraints. To be clear, this is not some kind of dualism of the spiritual/material, but rather a portrait of God working *within* creation through grace, allowing humans to freely respond.

One might liken this phenomenon to having God on a rope. If one can imagine the moment of justification as the moment when the Christian becomes connected to God via a rope (in this allegory, the rope signifies grace). Through the Holy Spirit's urgings, the Christian can draw the rope inward, shortening the distance between the Christian

---

I also do not wish to convey the idea that the human has become something other than fully human. It is my argument, then, that the word *overcome* expresses the truth that, by the human will and the grace of God, one is not bound by genetic proclivities.

[102]To be clear, if it were the case that genes totally determined actions without free will, rather than merely predisposing actions, it would skew what we think about sin. Yet, as discussed in the previous chapter, there is no reason to believe that genes do totally determine actions.

and God (and ultimately entire sanctification). While the rope is always connected to the individual, the Christian can let out the rope—freely moving away from entire sanctification and rejecting grace—or take in the rope.[103] At some point, however, the Christian is so closely connected to God that the rope does not have to connect the two. A Christian's works, then, is always a *response* to grace, where entire sanctification is unable to be earned when divorced from God's activity.

An analogy of a kite is another word picture that can be helpful to the idea of how God's grace working with human freedom could allow someone to overcome genetic urges and constraints. In order to fly a kite, one is incredibly active: putting together the kite, clearing space for running and flying, checking for imperfections in the tail or kite itself and so forth. Yet, flying a kite is impossible to do without wind. In this scenario, the wind is the active agent that allows an individual to fly the kite. All the good works of a Christian are only intelligible if they are working with the wind. Likewise, the works of a Christian are not enough if God is not active in the agent through his free grace. Thus, a Wesleyan concept of Christian perfection can be compatible with current knowledge of sociobiology if we are to assume that God works with the Christian not to replace or work in parallel with genetic material, but rather to work within the total human context by providing the grace by which a Christian can overcome negative genetic constraints.

Wesley's holistic anthropology was also compatible with sociobi-ology.[104] He strove earnestly to reject dualism of the body/soul through his understanding of *theosis* and the restoration of the *imago Dei*. At first glance, it might seem as though Wesley distinguished between two dimensions of life, seeing humans as "embodied souls/spirits."[105] Yet

---

[103]I have found this analogy, while unhelpful to some, to also be very helpful, especially when trying to explain how a Christian might be able to "throw away" their salvation, but not "lose it." For instance, after justifying grace, the individual is connected with God. They may let out the rope to an extended distance, but are still attached, and thus not cannot "lose" their salvation (as one might lose a wallet or car keys). At some point, however, the individual might choose to untie themselves from God and throw the rope aside.

[104]I took the idea of Wesley's "holistic anthropology" from Maddox's *Responsible Grace*. I also am indebted to Maddox for his thoughts within that section. See Randy L. Maddox, *Responsible Grace: John Wesley's Practical Theology* (Nashville: Kingswood Books, 1994), 70-72.

[105]Today, the language of *embodied spirits* is used to make a holistic rather than dualistic point. See ibid., 71.

this was the prevailing theme of his day—not to mention the fact that sociobiology, neuroscience and other salient sources of knowledge were not yet extant. Consequently, Wesley must be viewed in context, all the while noting Wesley's uncommon pushback against the dualism of his day. Despite the dualist's desire to separate the soul from the body, Wesley approached the issue by asking questions about how the soul was located *within* the body.[106] This distinction should not be considered slight; for Wesley thought that God preserved some kind of mysterious workings between the functions of the soul and brain.[107] Again, considering the time at which Wesley was situated in history, such an idea implies a significant connection of the soul to the body. As Maddox mentions, "[Wesley] rejected both a materialist reduction of this relationship and . . . [a] reduction of all creaturely action to God's immediate causation. For Wesley, both of these extremes discount the divine gift of liberty present in the human soul, and undercut the *responsibility* correlated with that *gracious* gift."[108] Though Wesley may have believed the soul would survive after bodily death, he also clearly thought the soul would be reembodied at the resurrection.[109] As Maddox rightly articulates, "So integral was the embodiment of the soul to him, in fact, that he occasionally advanced a distinction between 'body' and 'flesh and blood' which allowed him to assert that the soul was embodied by an 'ethereal body' even in its intermediate state. In short, while Wesley viewed the body and the soul as distinct realities, he did not view them as inappropriately conjoined."[110] Maddox goes on to say,

> Some Greek portrayals of the body/soul relationship assigned the body a primitive, if not actively antagonistic, impact on spiritual life. By contrast, the Bible presents the body as part of God's original good creation, and sin as a distortion of *every* dimension of human life. Wesley's direct comments on this point typically side with Scripture: he decried the philosophical contempt of the body; rejecting any claim that matter was the source of evil; and argued

---

[106]Ibid.
[107]Ibid.
[108]Ibid. Emphasis his.
[109]For multiple examples of this, see a list in ibid., 290n50.
[110]Ibid., 71.

that the biblical notion of sinful "flesh" did not refer to the body per se, but to the corruption of all dimensions of human nature.[111]

What is more, this care for both the body and the spirit was at the heart of Wesley's understanding of *theosis* and was part of his holistic anthropology.

To give greater detail to Wesley's anthropology, Maddox says that Wesley captured the Eastern Orthodox distinction between the *image of God* and the *likeness of God*: "The proper enduring orientation of these affections would constitute the Christian tempers [or inward holiness] which is the Likeness of God."[112] Maddox notes, "Wesley consistently identified inward holiness with Christian tempers."[113] These tempers would be the notion of being created in the image of God and living in God's likeness.[114] He finally ends his thoughts by noting that "overall, allowing for some dualistic influences, it seems fair to say that Wesley's two-dimensional anthropology did not degenerate into a strong metaphysical or ethical dualism. He sought, in his basic anthropological convictions, to emulate the holism of biblical teachings."[115] It is this emphasis that caused Wesley not to divorce the earthly from the divine.

Along with addressing concerns about Wesley's anthropology, questions of agency might arise in regard to how a Christian actually overcomes the biological. Is God the causal agent in human Christian perfection or does human action play a central role? The answer may be similar to that which Paul proposes in Philippians 2:12-13, where he encourages those at the church in Philippi to "work out your own salvation with fear and trembling; for it is God who is at work in you, enabling you both to will and to work for his good pleasure." If we hold that salvation for Wesley was meant to encompass Christian perfection in this life, Wesley's commentary, called *Notes of the New Testament*, proves

---

[111]Ibid., 72.

[112]Ibid., 7.

[113]Ibid., 289n35.

[114]Ibid., 73.

[115]Ibid., 72. It should be noted that Maddox continues with this caveat: "At the same time, it must be admitted that his valuation of bodiliness was not as positive, and his conception of the interrelationship of body and soul was not as integral and dynamic, as present theologians might desire." Yet, the progressive nature of Wesley's views for his time would suggest that he might be very sympathetic to the possibility that our bodies might influence our behavior.

extremely revealing. In it, he articulates that "not for any merit of yours. Yet his influences are not to supersede, but to encourage, our own efforts. *Work out your own salvation*—Here is our duty. *For it is God that worketh in you*—Here is our encouragement. And O, what a glorious encouragement, to have the arm of Omnipotence stretched out for our support and our succour!"[116] According to Wesley, both God and humanity play a critical role in Christian perfection: God acting through grace and humans acting through volition of the will.

If humans are predisposed by their biological makeup, but do not necessarily have to act on those predispositions, then the idea of *theosis* can illuminate the intersection of sociobiology and Christian perfection. *Theosis* for Wesley is more pragmatic than it is esoteric, focused rather on what he thought potential for this life, as mentioned before.[117] And as sociobiologists have in no uncertain terms proclaimed, genetics *also* matter for this life.[118] Because Wesley held the Eastern Orthodox understanding of salvation to be the "the renewal of our souls after the image of God,"[119] this renewal could lead the Christian working toward salvation to overcome genes and achieve Christian perfection. As Randy Maddox states in *Responsible Grace*,

> For Wesley, then, the Spirit's work of sanctification was not merely a forensic declaration of how God will treat us (regardless of what we are in reality). Neither was it a matter of directly infusing virtues in Christian lives. It was a process of character-formation that is made possible by a restored participation of fallen humanity in the Divine life and power.[120]

This restoration of the *imago Dei* gives both the idea that human action is necessary along with divine intervention.

When considering the ways in which Wesley's understanding of perfection might coincide with what we know about the human biological

---

[116]See John Wesley, *Explanatory Notes upon the New Testament*, vol. 2, *Romans to Revelation* (Kansas City, MO: Beacon Hill Press, 1981). Emphasis his. See also sermon 85, "On Working Out Our Own Salvation," in John Wesley, *The Works of John Wesley*, 6:506-13.

[117]Christensen, "Theosis and Sanctification," 80.

[118]If time and space had allowed, an interesting subject matter worth exploring would be the intersection of *theosis*, genetics and the "life to come." For instance, lions do not have the genetic proclivity to lie down with lambs.

[119]Maddox, *Responsible Grace*, 67.

[120]Ibid., 122.

makeup, it is important to note that he provided a few caveats concerning perfection, which help us frame how Christian perfection might work under the recent knowledge of our biological makeup. In "A Plain Account of Christian Perfection," Wesley cites eleven ways perfection functions in this life, but in a unique way.[121] Below are the most pertinent statements (in italics) along with my commentary:

(4) It is not absolute. Absolute perfection belongs not to man, nor to angels, but to God alone. This "absolute" that Wesley mentions seems to refer to humankind unaided by any grace from God.

(5) It does not make a man infallible: None is infallible, while he remains in the body. Wesley was concerned with critics confusing Christian perfection with "never making mistakes." This might support the claim that even while in a state of Christian perfection, a Christian is not practicing "perfectionism" and is still influenced by determined factors such as genes. As noted earlier in this chapter, Wesley's use of the phrase *while he remains in the body* should be understood in light of his holistic anthropology and his particular context.

(6) Is it sinless? It is not worth while to contend for a term. It is "salvation from sin." This statement would bolster the idea that, just as original sin remains but is overcome by prevenient grace, so those who are in a state of Christian perfection are not replacing genes but rather overcoming them.

(8) It is improvable. It is so far from lying in an indivisible point, from being incapable of increase, that one perfected in love may grow in grace far swifter than he did before. Due to the fact that Christian perfection was meant to be a restoration of the *imago Dei* in this life (not the life to come), it seems only logical that one could grow in grace even after reaching a state of perfection.

(9) It is amissible, capable of being lost; of which we have numerous instances. But we were not thoroughly convinced of this till five or six years ago. Wesley provides a statement and an anecdote that shows the ability of one to lose Christian perfection. Much like the other works of grace (justification, regeneration etc.), these states are not permanent. Rather, they perpetually hinge on the free will and actions of the Christian. By stating this, Wesley shows how Christian perfection is both a product of free will to be lost, *and* a grace to be freely given by God. Furthermore, what is unique about this statement is that it leaves open the possibility that biological proclivities might still hold

---

[121]All of these statements are taken from "A Plain Account of Christian Perfection" found in Wesley, *The Works of John Wesley*, 11:441-42. For a complete list, see appendix two.

sway over a Christian. Thus, genes are not being replaced by grace but are rather overcome through grace. Statements ten and eleven also support this theory, claiming that perfection is a gradual process that leads up to an instantaneous moment of grace. Again, Wesley is trying to convey the idea that perfection is both human action and God ordained. And it is the human action that can succumb to the selfish biological tendencies still remaining in the human-Christian.[122]

It is worth mentioning that throughout the list, Wesley refers to justifying grace (along with other forms of grace) numerous times. A conclusion that can be drawn from this is that Wesley saw Christian perfection as yet another form of grace—maybe even the last grace in a Christian's life. Consequently, it would seem that this last grace of Christian perfection would be a reckoning of the fact that humans were originally made in a state of perfection and had fallen away.[123] As Outler suggests, "It is almost as if Wesley had read ἀγάπη [love] in the place of the Clementine γνῶσις [cognition], and then had turned the Eastern notion of a vertical scale of perfection into a genetic scale of development within historical existence."[124]

One clear concept can be gathered from what we know of how Wesley thought of Christian perfection: it cannot be attained merely through a matter of the will. If reaching a state of perfection is just a matter of the will, biological constraints could make the achievement of such impossible for some, which would call the whole concept into question. So, for Wesley, the idea of divine grace is combined with human freedom, making the notion of perfection more probable; and it is owing to this combination that the idea that humans were created in a perfect state became the prevailing perspective in Wesleyan traditions.[125] In "A Farther Appeal to Men of Reason and Religion," Wesley states, "By salvation I mean, not

---

[122]Certainly this concept of "perfection" is complicated and not without argument and interpretation—both in Wesley's day and ours. My hope in working through his statements is that the reader might see that evolutionary biology and the idea of Wesleyan Christian perfection are not incompatible.

[123]See the sermon titled "The Image of God," found in John Wesley, *Sermons 115–151*, ed. Albert C. Outler, The Bicentennial Edition of the Works of John Wesley (Nashville: Abingdon Press, 1987), 290-303. For commentary on this, see also Maddox, *Responsible Grace*, 67.

[124]Wesley, *John Wesley: A Representative Collection of His Writings*, 31.

[125]Maddox, *Responsible Grace*, 65.

barely, according to the vulgar notion, deliverance from hell, or going to heaven; but a present deliverance from sin, *a restoration of the soul to its primitive health, its original purity; a recovery of the divine nature; the renewal of our souls after the image of God, in righteousness and true holiness, in justice, mercy, and truth.*"[126] This restoration in the context of "salvation" cannot be assumed to happen without some kind of notion of grace. Thus, the Christian, by the grace of God, overcomes biological predispositions to be completely restored to the holistic perfection of prefallen humanity.[127] It is only when the action of sin is dealt with by humans that holiness can be attained.[128] This is why Wesley was so interested in constraining behavior with his accountability groups (see chapter six). As Wesley describes Adam's prefallen state in sermon 60, "The General Deliverance," we see a positive statement of human interaction with divine grace: "[Adam] was a creature capable of God, capable of knowing, loving and obeying his Creator. And in fact he did know God, did unfeignedly love and uniformly obey him. This was the supreme perfection of man, as it is of all intelligent beings—the continually seeing and loving and obeying the Father of the spirits of all flesh."[129]

5.3 CONCLUSION

In this chapter, I sought to articulate how one might understand Wesleyan holiness against a backdrop of evolutionary biology. John Wesley was driven to encourage his followers to become holy people. This quest for holiness was undergirded by his theological understanding of original sin and Christian perfection, leading directly to the shape and development of Methodist thought. Specifically, Wesley's doctrine of Christian perfection points toward the process through which divine grace aids the individual to "overcome" biology, not to replace her or his biological

---

[126]Wesley, *The Works of John Wesley*, 8:47. Emphasis mine.

[127]It should also be noted that prefallen humans—though perfect as they were—still had the free will to fall from grace.

[128]Wesley, *The Works of John Wesley*, 11:38.

[129]Wesley, *Sermons 34–70*, 439. Again, without completely adopting Eastern ideas of the fall, the Augustinian framework makes it a bit more difficult to understand contemporary research in sociobiology. To the degree that Wesley articulates the Eastern understanding of the fall and redemption, he ends up being more compatible with sociobiology. And to the degree Wesley follows the Augustinian concept of original sin, he remains more in tension with sociobiology.

makeup. This grace-dependent overcoming enables Christians not to be constrained by biological tendencies toward selfishness.

Wesley thought that Christians had free will to choose good action, despite the many constraints placed on the individual (whether biological constraints or constraints from the fall). Yet he also saw God as being intimately involved in an individual's salvation journey. One can see this in the way he brings forth prevenient grace to help mitigate the potency of original sin. One can also notice the concept of grace and free will at work in concert in Wesleyan doctrines of justification, regeneration and the process of sanctification. Thus, it would follow that Wesley viewed Christian perfection as both an endeavor of free will and divine grace.

With the Wesleyan understanding of the human condition being a combination of free will and divine grace, Christians in a state of perfection have not replaced their genes, nor is the individual living with the parallel of dualism where the body and the soul are functioning side by side as two distinct and separable entities. Instead, the Christian in a state of perfection has overcome her or his genes, not being totally constrained by the proclivities of her or his own biology, rather having her or his whole desire focused on God through grace in Christ. This overcoming does not mean that free will does not exist for the perfected Christian. On the contrary, the individual has the ability to retreat through her or his own volition, allowing her or his biological selfish tendencies to have dominance again. In this way, this concept of "overcoming" maintains a holism of body and soul, displaying how the Christian can remain completely in their body yet enacted upon by the divine grace of Christian perfection.

It is noteworthy that Wesley often used the phrase "Not as though I had already attained" to accompany discussions on Christian perfection.[130] While he believed that Christian perfection could be attained, his caveat always ensured that a person was able to grow *more* in Christian perfection.[131] In sermon 120, "The Wedding Garment," Wesley says, "In a

---

[130]For some examples, see John Wesley, *The Methodist Societies: History, Nature, and Design*, ed. Rupert E. Davies, The Bicentennial Edition of the Works of John Wesley (Nashville: Abingdon Press, 1989), 30-46. See also Wesley, *Sermons 34–70*, 96-124.

[131]Maddox, *Responsible Grace*, 190.

word, holiness is having 'the mind that was in Christ' and 'the walking as he walked.'"[132] Of the same subject he also proclaimed, "The sum of Christian perfection is all compromised in that one word, love."[133] Wesley purposely defines holiness to show that there is always room for progress—even beyond perfection.[134] As Wesley continues, "So that how much soever any man has attained, or in how high a degree soever he is perfect, he hath still need to 'grow in grace' and daily to advance in the knowledge and love of God his Saviour."[135] This progression after entire sanctification is but another example that Christian perfection is only attainable through an overcoming of biological constraints—grace working with the confines of creation. Consequently it is John Wesley's theological understanding of the human person, influenced by both original sin and works of grace enabling perfection in this life, that lay the foundation for his accountability groups that move individuals toward a lifestyle of selflessness.

---

[132]John Wesley, *The Works of John Wesley*, 7:317.
[133]Ibid., 6:413.
[134]Moore, "Development in Wesley's Thought on Sanctification and Perfection," 35.
[135]Wesley, *John Wesley: A Representative Collection of His Writings*, 258.

6

HOW WESLEY NURTURED
ALTRUISM DESPITE BIOLOGICAL
CONSTRAINTS

◆

In previous chapters, I have shown how biology influences the human condition and explained the inability of sociobiologists to fully make sense of altruistic actions. I have also elucidated the natural constraints on an individual's behavior, both genetic and environmental, and how these constraints impact human freedom and responsibility. It would stand to reason, then, that if the biological human condition rests somewhere between pure egoism and altruism, we would do well to investigate what environmental constraints might push individuals closer to altruism and how one might foster those constraints. Therefore, we will consider in this chapter how John Wesley's small groups encouraged people to be more altruistic; what is more, though it would be anachronistic to say that Wesley knew about the genetics of the biological human condition, he placed people in groups for both the practical reasons of organization and the theological reasons of engendering holiness. In fact, as I will argue, it is through what I will call Wesley's *world of constraints* that he placed on his early followers that Wesley strengthened their altruistic tendencies and mitigated their egoistic tendencies.

Such a claim will require us first to look at the specific structure and organization of the early Methodist groups, which first requires us to define some terminology and to consider the culture of the early Methodists before working through the reasons behind such groups. Thus, we will walk through a brief history of how the movement came about as well as the specific accountability structure that included a host of environmental constraints put forth by Wesley. Accordingly, it is important to account for Wesley's theological understanding of holiness (which is broken down into *inward* and *outward* holiness). For Wesley, theology was practical, and his sermons became tools for spiritual formation.[1] Understanding Wesley's view will provide us with a glimpse of the foundation and assumptions with which Wesley was working in order to better understand how his groups nurtured altruism. His core theological concepts included a nuanced view of the doctrine of original sin that was coupled with prevenient grace,[2] allowing us to draw a direct parallel between his theological concepts and current sociobiological discoveries that point to human nature as being not totally egoistic but bearing altruistic behavior and potential. We will also discuss Wesley's theology as the spiritual dimension to the early movement. It is important to note that Wesley saw the movement as having a spiritual source that included religious experiences. All of these theological ideas laid the groundwork for Wesley to assemble a world of constraints that would nurture altruism by way of holiness.

Although altruism is the product of a particular community, it goes beyond the constraints of that community, calling for a limitless concern for all human beings. Thus, Christians living with such altruism cultivate a way of life through hospitality and charity that naturally leads to helping others.[3] In Luke 6:33, Christ claims, "If you do good to those who do good to you, what credit is that to you? For even sinners do the

---

[1]John Wesley, *Sermons 1–33*, ed. Albert C. Outler, The Bicentennial Edition of the Works of John Wesley (Nashville: Abingdon Press, 1984), xiii.

[2]In response to how these ideas were lived out in reality (as well as other soteriological questions), Wesley states in his sermon 43, "The Scripture Way of Salvation," "The salvation which is here spoken of is not what is frequently understood by that word, the going to heaven, eternal happiness. . . . It is not a blessing which lies on the other side of death. . . . It is a present thing. . . . [It] might be extended to the entire work of God from the first dawning of grace in the soul till it is consummated in glory." See John Wesley, *Sermons 34–70*, ed. Albert C. Outler, The Bicentennial Edition of the Works of John Wesley (Nashville: Abingdon Press, 1985), 156. The major focus for salvation, for Wesley, was transforming the here and now. (I am indebted to conversations with Darrell Moore and to his unpublished essay for a number of ideas in this chapter.) See Darrell Moore, "Classical Wesleyanism" (unpublished paper, 2011).

[3]Christine D. Pohl, *Making Room: Recovering Hospitality as a Christian Tradition* (Grand Rapids: Eerdmans, 1999), 35.

same." Perhaps we would do well here to understand Christ's words as a call to rise above the simple genetic reciprocity that might be engraved in our DNA. To this end, Wesley capitalized on the human condition, a natural phenomenon that rests somewhere between the extremes of self-ishness and selflessness. Despite the fact that Wesley was pre-Darwinian, he intuitively understood humankind. With his intense drive to find ways both to care for the poor and to dwell in community with them, Wesley instituted a system to help alleviate their plight and to encourage them toward any natural inclinations of altruism: his famous bands and classes. Besides the Moravians (also known as the Unitas Fratrum), there were no other contemporary examples of the kind of small-group re-newal movement happening at the time.[4]

Many of the individuals whom Wesley was shepherding needed structure because many were troubled with alcoholism or other dysfunc-tional and self-destructive lifestyles.[5] Consequently, Wesley early on found it necessary to develop systems of checks to help monitor and en-courage good behavior, revealing himself to be a pragmatist. However, he was also relational, finding ways of connecting to the common person, even when understanding a person's lifestyle made him uncomfortable. Wesley particularly wanted to see his new structured groups preserved and made available to people for traditional Sunday services, and was slower to see how this could be meaningfully adapted. In his Saturday, March 31, 1739, journal entry concerning his first encounter with George Whitefield's "open air" preaching style, Wesley says, "In the evening I reached Bristol, and met Mr. Whitefield there. I could scarce reconcile myself at first to this strange way of preaching in the fields, of which he set me an example on Sunday; having been all my life (till very lately) so

---

[4]John Wesley, *The Methodist Societies: History, Nature, and Design*, ed. Rupert E. Davies, The Bi-centennial Edition of the Works of John Wesley (Nashville: Abingdon Press, 1989), 3. In fact, as J. Wesley Bready notes, the "evangelical revival did more to transfigure the moral character of the general populace, than any other movement British history can record." See Bready, *England: Before and After Wesley* (London: Hodder and Stoughton, 1938), 327.
[5]Idea taken from Moore, "Classical Wesleyanism," 13. Mildred Bangs Wynkoop speaks clearly to the kind of problems people would deal with in the early Methodist movement. She discusses the "heart" issues that resulted in outward acts of self-destruction. She says, "The very nature of sin is love's perversion which makes the self the object of its own dedication." See Wynkoop, *A Theology of Love: The Dynamic of Wesleyanism* (Kansas City, MO: Beacon Hill Press, 1972), 18.

tenacious of every point relating to decency and order, that I should have thought the saving of souls almost a sin, if it had not been done in a church."[6] Yet Wesley himself had changed, and he was willing to work outside churches, with the down-and-out of society, in order to move people toward holiness. To care for the poor, Wesley was willing to be flexible on some of the nonessential issues within his own belief system. Wesley then caught the pragmatist fire to reach people wherever they were and took his open-air preaching further still. He says on Monday, April 2, 1739, that "at four in the afternoon, I submitted to be more vile, and proclaimed in the highways the glad tidings of salvation, speaking from a little eminence in a ground adjoining to the city, to about three thousand people."[7] As Howard Snyder notes in his book *The Radical Wesley*, Wesley—in characteristic form—"immediately began to organize" and added that "from the beginning [the Wesleyan Revival] was a movement largely for and among the poor, those whom 'gentlemen' and 'ladies' looked on simply as part of the machinery of the new industrial system."[8]

6.1 WESLEY'S STRUCTURE AND ORGANIZATION

It might first be helpful to very quickly walk through the inception of the early Methodist movement as well as discussing its leader, John Wesley, and his personal experiences. These events helped play a key role in the development of Wesley's highly organized and structured groups.

The early Methodist societies can be traced back to the 1729 Oxford prayer and study meetings of four men: John and Charles Wesley, William Morgan (a commoner of Christ Church), and Robert Kirkham of Merton College. Over several years, the group grew to later include George Whitefield in 1735.[9] This "Holy Club," as it became known, was where Wesley first saw the benefits of accountability and regular meeting. It was also where he had a significant religious experience that led him to reform his thinking. Interestingly, both the names Holy Club and

---

[6]John Wesley, *The Works of John Wesley*, ed. Thomas Jackson and Albert C. Outler, 14 vols. (Grand Rapids: Zondervan, 1958), 1:185.
[7]Ibid.
[8]Howard A. Snyder, *The Radical Wesley & Patterns for Church Renewal* (Downers Grove, IL: Inter-Varsity Press, 1980), 33. Based in part on the author's Notre Dame thesis.
[9]Wesley, *The Works of John Wesley*, 8:348.

Methodists started out as pejoratives. In *A Short History of Methodism*, Wesley states, "It is not easy to reckon up the various accounts which have been given of the people called Methodists; very many of them as far remote from truth as that given by the good gentleman in Ireland: '*Methodists!* Ay, they are the people who place all religion in *wearing long beards*.'"[10] More precisely, Wesley notes in the same text, "The exact regularity of their lives, as well as studies, occasioned a young gentleman of Christ Church to say, 'Here is a new set of Methodists sprung up;' alluding to some ancient Physicians who were so called. The name was new and quaint; so it took immediately, and the Methodists were known all over the University."[11] Thus the early group took on the derogatory term with pride—as a badge of honor.

As noted by Rupert Davies, Wesley's early Methodists fall somewhere between a church and a sect and can be labeled a "Christian communion."[12] For example, one of the main functions of a church is to protect the catholic tradition of sacraments and worship while a sect seeks to separate itself from the life of the church.[13] Yet Wesley tried to stay within the Church of England and had no real desire to break away. He thought that the best location for renewal groups would be *within* the Church of England.[14] We know about Wesley's desire to remain connected to the larger church because he believed that a society ought to acknowledge "the truths proclaimed by the universal church" and ought not "wish to separate from it," but instead "cultivate, by means of sacrament and

---

[10]Ibid., 347.

[11]Ibid., 348.

[12]See Wesley, *The Methodist Societies*, 2-3. Here, Davies is critiquing the pervasive sociological treaties that categorize new religious groups into a dichotomy of either sect or church. The most influential literature is the 1912 work by Ernst Troeltsch, *Die Soziallehre der Christlichen Kirchen und Gruppen, Gesammelte Schriften* (Tübingen: J. C. B. Mohr, 1912). For an English translation see Troeltsch, *The Social Teaching of the Christian Churches*, trans. Olive Wyon (Chicago: University of Chicago Press, 1981).

[13]Wesley, *The Methodist Societies*, 3.

[14]Trying to explain that his early Methodists were not separatists from the Church of England, Wesley states in *A Plain Account of the People Called Methodists* that he is not destroying the church, but building it up. He says, "Is this Christian fellowship there? Rather, are not the bulk of the parishioners a mere rope of sand [in other societies]? What Christian connexion [*sic*] is there between them? What intercourse in spiritual things? What watching over each other's souls? What bearing of one another's burdens? What a mere jest is it then, to talk so gravely of destroying what never was!" Wesley, *The Works of John Wesley*, 8:251-52.

fellowship, the type of inward holiness, which too great an objectivity can easily neglect and of which the church needs constantly to be reminded."[15]

Wesley always planned his class meeting so that there was no conflict with regular Anglican services, and continually urged his people to be faithful to the church.[16] He also insisted that the early Methodists diligently attend all the services of the established church, where they were expected to be among the most devoted members.[17] Methodist children were not excluded from this expectation. The Methodists were to be baptized by parish vicars—and no meeting could be held in a Methodist hall while an Anglican Church service was commencing.[18] Wesley also notes, in his *Short History of Methodism,* that the initial participants of the movement were "all zealous members of the Church of England; not only tenacious of all the doctrines, so far as they knew them, but of all her discipline, to the minutest circumstance."[19] He adds to the significance of the Church later in the same work by noting, "At present, those who remain [Methodist] are mostly Church-of-England men. They love her Articles, her Homilies, her Liturgy, her discipline, and unwillingly vary from it in any instance."[20] It seems as though Wesley's words here are imbued with a sense of pride in his desire continually to have a close connection with the Church of England.

The title of chapter fourteen in Henry Racks's biography of Wesley, *Reasonable Enthusiast,* describes Wesley's attitude toward the Church of England: "I Live and Die in the Church of England."[21] It was not until the time of the Revolutionary War that the Methodists began to break away from the Church of England—which happened first in America, for obvious political and social reasons having to do with America's opposition

---

[15]Wesley and Davies, *The Methodist Societies,* 3.

[16]Moore, "Classical Wesleyanism," 18.

[17]Maximin Piette and Joseph Bernard Howard, *John Wesley in the Evolution of Protestantism* (New York: Sheed & Ward, 1937), 466.

[18]Ibid.

[19]Wesley, *The Works of John Wesley,* 8:348.

[20]Ibid., 350.

[21]Henry D. Rack, *Reasonable Enthusiast: John Wesley and the Rise of Methodism,* 2nd ed. (Nashville: Abingdon Press, 1993), 489-534. Much of the following ideas, broadly speaking, are influenced by Rack's account. For this quotation in full, see John Wesley's letter to Henry Moore on May 6, 1788, where he said, "I am a Church-of-England man; and, as I said fifty years ago so I say still, in the Church I will live and die, unless I'm thrust out." See Robert Southey, *The Life of Wesley: And the Rise and Progress of Methodism* (London: Frederick Warne, 1889), 523.

to England; mainly, American Methodists did not want to be damaged by Loyalist associations. Yet, Wesley never approved of the war and did not see the Methodist break from the Church of England as a positive thing; in fact, up until his death, he remained faithful to the Church of England. After his death, however, the Methodists residing in England eventually broke away from the Church of England as well. Nevertheless, though Wesley's desire to stay with the Church of England was not ultimately sustained, his influence on both the Methodists and the Church of England can still be felt even today.

6.1.1 Wesley's highly structured groups

A major goal for Wesley was to develop a personal religious experience within the context of supportive accountability groups.[22] As Howard Snyder says, "John Wesley saw that new wine must be put into new wineskins. So the story of Wesley's life and ministry is the story of creating and adapting structures to serve the burgeoning revival movement."[23] Wesley looked at the new Methodists and pursued a structure that would buttress the regimented lifestyle of the fledgling group. The system that developed proved wrong those who questioned whether a church movement could be built among the poor and uneducated.[24] In pursuit of such a system, Wesley sought the guidance of Anglican and non-Anglican individuals alike. Peter Böhler, a Moravian friend, convinced Wesley that small groups, known as "bands," were necessary for accountability and health.[25] In turn, and to provide care to the many bands that would spring up, Wesley made it a point to methodically reach countless followers through lay ministers working with each band.[26]

Before his Aldersgate conversion, Wesley undoubtedly noticed his own propensity to waver back and forth in spiritual and temporal disciplines. He formed a small group of close individuals to help him stay steadfast in his faith (the Holy Club). He was also quick to point out that

---

[22]Paul Wesley Chilcote, *Recapturing the Wesleys' Vision: An Introduction to the Faith of John and Charles Wesley* (Downers Grove, IL: InterVarsity Press, 2004), 45.

[23]Snyder, *The Radical Wesley & Patterns for Church Renewal*, 53.

[24]Ibid.

[25]David Lowes Watson, *The Early Methodist Class Meeting: Its Origins and Significance* (Nashville: Discipleship Resources, 1985), 80-81.

[26]Snyder, *The Radical Wesley & Patterns for Church Renewal*, 53.

although the Christian faith is personal, it is not private; Christianity is rather a "social religion," and to relegate it to individual faith without accountability "is indeed to destroy it."[27] In this way, we find that the early Holy Club that John Wesley belonged to with his brother, Charles, was focused not only on cultivating the general knowledge and inward piety of those in the club, but looked outward as well. It was through the direction of this group that Wesley first recognized that the social work of caring for the poor must be an inseparable part of Christian living.[28] In fact, he was one of the first not only to understand the poor as recipients of the alms and charity, but also to help them engage in acts of charity themselves by encouraging them to visit the sick, imprisoned and otherwise burdened people in their communities.[29] It is thus within these early classes and bands that we see the genesis of the process of nurturing altruism among individuals—regardless of the member's social or financial status.

These Moravian-influenced bands were one of the first groups that Wesley initiated.[30] Very different from larger meetings (either within the Church of England or as part of the Methodist renewal group), the bands were not focused on church discipline but were instead—having been formed of five to ten individuals (same gender and marital status)— focused on helping new converts progress in faith and actions.[31] These meetings were not intended to break individuals away from their home church, but rather to "call its own members within the larger church to a special personal commitment which respects the commitment of others."[32] Bands differed from classes (larger Methodist meetings) in that they were restricted to people who had a measure of assurance of the remission of sins.[33] Whereas the band's focus was on the *spiritual growth* of the individual, the class meetings focused on renewing *discipline*

---

[27]Kenneth J. Collins, *The Scripture Way of Salvation: The Heart of John Wesley's Theology* (Nashville: Abingdon Press, 1997), 61-62. See also sermon 24 from Wesley, *Sermons 1–33*, 531.

[28]Manfred Marquardt, *John Wesley's Social Ethics: Praxis and Principles* (Nashville: Abingdon Press, 1992), 23.

[29]Ibid., 27.

[30]Snyder, *The Radical Wesley & Patterns for Church Renewal*, 35-36.

[31]Ibid., 60.

[32]Wesley, *The Methodist Societies*, 3.

[33]Snyder, *The Radical Wesley & Patterns for Church Renewal*, 60.

within the individual.[34] Yet both band and class worked together to form the structure of accountability in Wesley's organized movement. In fact, Wesley refused to minister in any place where he could not follow up with organized groups with structured leadership[35]—an organization that became his hallmark, and a necessary part of the early movement.

Given the nature of the small groups, band members were held to a high standard of living and an expectation of holistic growth that was spiritual, intellectual and encouraged abstention from certain physical vices. Band members were to abstain from doing evil, to be enthusiastic in good works, "including giving to the poor, and to use all the means of grace."[36] As Snyder notes,

> Understandably, with this kind of rigor fewer bands were organized than classes. Judging from the number of band and class tickets printed, it would appear that about twenty percent of the Methodist people met in bands, whereas all were class members. Since the bands averaged about six members and the classes about twelve, this means there were probably about two or three classes for every band.[37]

Wesley also initiated a "Select Society" particularly for those who were making progress toward both inward and outward holiness. We find a powerful example of Wesley developing the philosophies of his groups in his sermon 61, "The Mystery of Iniquity," wherein he describes how the early believers held all things in common:

> "How came they to act thus, to have all things in common, seeing we do not read of any positive command to do this?" I answer, There needed no outward command: the command was written on their hearts. It naturally and necessarily resulted from the degree of love which they enjoyed. Observe! "They were of one heart, and of one soul": And not so much as one (so the words run) said, (they could not, while their hearts so overflowed with love,) "that any of the things which he possessed was his own." And wheresoever the same cause shall prevail, the same effect will naturally follow.[38]

---

[34]Ibid., 38.

[35]Sydney George Dimond, *The Psychology of the Methodist Revival: An Empirical and Descriptive Study* (London: Oxford University Press, 1926), 112.

[36]Snyder, *The Radical Wesley & Patterns for Church Renewal*, 59.

[37]Ibid., 60.

[38]Wesley, *The Works of John Wesley*, 6:256. See also Snyder, *The Radical Wesley & Patterns for Church Renewal*, 174n27.

For Wesley, the proper Christian life was necessarily both "profoundly personal and essentially social" and significantly structured.[39] What is more, out of Wesley's ideas we are able to cull three rules that were attached to his groups:[40] everything that is spoken in the select society remains private, even the members' identity; every member must be in submission to the minister; every member (until "all things are held in common") will give a once-a-week gift, all she or he can give toward the common group. It is thus in his description of "all things common" that we are able to identify Wesley's ideal of a true community of goods among those who were the most committed members of the early Methodist groups.[41] This unique approach, which stems from his idea of *inward* holiness, leads to *outward* actions. Therefore, the health and sustainability of these groups was tied to the selfless imperative that members of the small groups foster inward holiness, or they would be unable to attain outward holiness.

By 1738, Wesley had set up a complex and thriving system for accountability and spiritual growth filled with bands, classes and societies.[42] The discipline that Wesley cultivated and planted gave way to a growing body of sincerely changed followers. In 1768, after thirty years of structured bands and classes, the Methodists had a total of forty circuits and over twenty-seven thousand members. In 1778, there were sixty circuits and over sixty-six thousand members. By 1798, the Methodists had grown to nearly 150 circuits and over one hundred thousand people in membership—a membership in highly structured accountability groups that required regular attendance and that answered still to the Anglican Church.[43] While Methodist societies were divided into classes and then bands, it is perhaps more helpful when thinking of the Methodist structure to say that societies were the sum of classes and bands, since one had to join a more intimate band and class before one was allowed

---

[39]Chilcote, *Recapturing the Wesleys' Vision*, 46.
[40]The following rules are drawn from Snyder, *The Radical Wesley & Patterns for Church Renewal*, 62.
[41]Ibid.
[42]Ibid., 53.
[43]E. Douglas Bebb, *Wesley: A Man with a Concern* (London: Epworth Press, 1950), 121-22. See also Snyder, *The Radical Wesley & Patterns for Church Renewal*, 54.

to take membership in a society.[44] After all, the purpose of the Methodist movement was small and focused accountability groups. The classes were effectively house churches, rather than "Sunday school" or other modes of more formal instruction. They met in the homes of members and the leaders functioned as pastors.[45]

During Wesley's early Methodist revival, as Stephen Long notes, religious movements were not focused on nurturing the natural inclinations of individuals because, at this point in time, they were divorcing practical ethics from theology, a mistake commonly made in the eighteenth century—one with which Wesley would have been all too familiar.[46] Instead, Wesley took the approach of tackling social issues *combined with* theology in his organized classes and bands in order to maintain a relationship between theology and praxis. In addition to other practical reasons for assembling bands and classes, Wesley had seen the lack of organization in George Whitefield's ministry, despite his eloquent preaching, and knew that organized and accountable groups would be the lifeblood to any movement of revival, and would foster among his groups not only inward holiness but also outward holiness.[47]

For an example of combined theology and praxis, Wesley encouraged the bands to develop selfless action as part of their lifestyle. According to Randy Maddox's *Responsible Grace*, Wesley's driving economic theme—both in and outside his classes and bands—was fourfold:[48] everything ultimately belongs to God, resources are placed in our care at God's discretion, God wants us to use those resources to meet our needs (food, clothing etc.) and then to meet the needs of others, to expend those resources on luxuries for ourselves while those around us remain in need is robbing God. In this way, Wesley's views here are practical, coinciding with the maxim in his sermon "The Use of Money" that we ought to earn all we can, save all we can and give all we can.[49]

---

[44]Ibid., 54.
[45]Ibid.
[46]D. Stephen Long, *John Wesley's Moral Theology: The Quest for God and Goodness* (Nashville: Kingswood Books, 2005), 203.
[47]Collins, *The Scripture Way of Salvation*, 160.
[48]Maddox, *Responsible Grace*, 244.
[49]Wesley, *Sermons 34–70*, 276–77.

Other similar movements, such as George Whitefield's revival groups, did not organize to either the extent or effect that the early Methodists did, and those revival groups subsequently fell away. Although there was a spiritual renewal happening throughout England at the time, it was Wesley's structure that produced lasting fruit. The following is a short conversation from Holland McTyeire, a late-nineteenth-century historian, that speaks to Wesley's organization:

> It was by this means (the formation of Societies) that we have been enabled to establish permanent and holy churches over the world. Mr. Wesley saw the necessity of this from the beginning. Mr. Whitefield, when he separated from Mr. Wesley, did not follow it. What was the consequence? The fruit of Mr. Whitefield's labors died with himself: Mr. Wesley's fruit remains, grows, increases, and multiplies exceedingly. Did Mr. Whitefield see his error? He did, but not till it was too late. His people, long unused to it, would not come under this discipline. Have I authority to say so? I have and you shall have it.
>
> Forty years age I traveled in Bradford, the Wilts Circuit, with Mr. John Pool. Himself told me the following anecdote. Mr. Pool was well known to Mr. Whitefield, and having met him one day, Whitefield accosted him in the following manner: "Well, John, art thou still a Wesleyan?" Pool replied, "Yes, sir, and I thank God that I have the privilege of being in connection with him, and one of his preachers." "John," said Whitefield, "thou art in thy right place. My Brother Wesley acted wisely—the souls that were awakened under his ministry he joined in class, and thus preserved the fruits of his labor. This I neglected, and my people are a rope of sand."[50]

While this high degree of structure stemmed from Wesley's own penchant for organization, it also highly benefited the newfound holiness movement.

---

[50]Holland Nimmons McTyeire, *A History of Methodism: Comprising a View of the Rise of This Revival of Spiritual Religion in the First Half of the Eighteenth Century, and of the Principal Agents by Whom It Was Promoted in Europe and America; With Some Account of the Doctrine and Polity of Episcopal Methodism in the United States, and the Means and Manner of Its Extension Down to A.D. 1884* (Nashville: Southern Methodist Publishing House, 1884), 204. See also D. Michael Henderson, *John Wesley's Class Meeting: A Model for Making Disciples* (Nappanee, IN: Evangel Pub. House, 1997), 30. It is interesting that George Whitefield was persuaded to part ways from Wesley due to his differing understanding of salvation. Wesley was inclined to believe in a "general salvation" where all have the potential to be saved, while Whitefield believed in "particular salvation" where only the elect are saved; a debate similar to that of contemporary Reformed and Armenian theologians. This disagreement between Wesley and Whitefield caused two groups of Methodists to emerge. See point eleven of *A Short History of Methodism* found in Wesley, *The Works of John Wesley*, 8:349.

6.1.2 Accountability structure within Wesley's groups

Before expanding on the world of constraints John Wesley placed on his followers, it is important to elucidate the accountability structure within early Methodist groups because it became the foundation for any constraints Wesley placed upon his followers. This accountability applied not only for the laity within the movement, but also for the lay leaders who functioned as quasiclergy for the Methodists. There was a high degree of expectation. In *A Plain Account of the People Called Methodists in a Letter to the Revd. Mr. Perronet*, Wesley mentions two major "businesses" of a leader:

> It is the business of a Leader (1) To see each person in his class, once a week at the least, in order to inquire how their souls prosper; to advise, reprove, comfort, or exhort, as occasion may require; to receive what they are willing to give toward the relief of the poor. (2) To meet the Minister and the Stewards of the Society, in order to inform the Minister of any that are sick, or of any that are disorderly and will not be reproved; to pay to the Stewards what they have received of their several classes in the week preceding.[51]

This list sheds extraordinary light on the kind of communities he was creating. These groups consisted of individuals under direct supervision and guidance, working toward both inward and outward holiness. Regiments like this helped shape a kind of people who, when called upon to give and sacrifice generously, would give as a natural extension of their lifestyle. The early Methodists were habitual in their practices and were all held accountable, from the lay to Wesley himself.

In sermon 107, "On God's Vineyard," the requirements of the classes were spelled out very clearly. Wesley states that a member is "placed in such a class as is convenient for him, where he spends about an hour in a week. And, the next quarter, if nothing is objected to him, he is admitted into the society: And therein he may continue as long as he continues to meet his brethren, and walks according to his profession."[52] We see here the ways in which Wesley's groups became, above all else, a

---

[51]John Wesley, *A Plain Account of the People Called Methodists in a Letter to the Revd. Mr. Perronet*, 2nd ed. (London: W. Strahan, 1749). See also Wesley, *The Works of John Wesley*, 8:253.

[52]Wesley, *The Works of John Wesley*, 7:209.

system of what Snyder calls "discipline-in-community."[53] The group discipline was not limited to leaders alone, but was highly structured for members as well and often included tangible care for the poor. This practice came with much accountability and expectation. We find this practice articulated again in Wesley's own words as he recalls a conversation with a member at the Bristol society:

> I was talking with several of the Society in Bristol concerning the means of paying the debts there, when one stood up and said, "Let every member of the Society give a penny a week till all are paid." Another answered, "But many of them are poor, and cannot afford to do it." "Then," said he, "put eleven of the poorest with me; and if they can give anything, well I will call on them weekly; and if they can give nothing, I will give for them as well as for myself. And each of you call on eleven of your neighbours weekly; receive what they give, and make up what is wanting." It was done. In a while, some of these informed me, they found such and such and one did not live as he ought. It struck me immediately, "This is the thing; the very thing we have wanted so long." I called together all the Leaders of the classes (so we used to term them and their companies), and desired that each would make a particular inquiry into the behaviour of those whom he saw weekly. They did so. Many disorderly walkers were detected. Some turned from the evil of their ways. Some were put away from us. Many saw it with fear, and rejoiced unto God with reverence.[54]

Wesley writes in his journal on Monday, February 15, 1742, that this was the accountability practice that should be instilled in the whole society. Here he says,

> Many met together to consult on a proper method for discharging the public debt; and it was at length agreed, 1. That every member of the society, who was able, should contribute a penny a week. 2. That the whole society should be divided into little companies or classes,—about twelve in each class. And, 3. That one person in each class should receive the contribution of the rest, and bring it in to the stewards, weekly.[55]

---

[53]Snyder, *The Radical Wesley & Patterns for Church Renewal*, 53.
[54]Wesley, *A Plain Account of the People Called Methodists in a Letter to the Revd. Mr. Perronet*. See also Wesley, *The Methodist Societies*, 260-61.
[55]Wesley, *The Works of John Wesley*, 1:357.

Thus, a weekly and highly organized system of accountability that focused on spiritual and practical aspects of the Christian life was established. One can see how serious discipline had to be practiced in the small bands where the leader intimately knew each individual.[56] The result of the accountability groups was that early Methodist people had the freedom of living in the grace of God within the context of close fellowship and committed community.[57]

6.1.3 Wesley's world of constraints

John Wesley created an environmental world of constraints that pushed people toward altruism. These constraints, which were the outcropping of his highly structured accountability groups, worked to suppress other biological tendencies that would have encouraged the individual to be more selfish. In a way, Wesley created an organized community that encouraged underlying natural tendencies toward altruism. Wesley's passion for change in people's lifestyle exuded in all that he did. With those who were ungenerous, he felt a strong need to encourage modifications in their behavior. One can easily track his encouragement in his sermons, especially those moments in which he employs fiery language when focusing on the seriousness of shifting one's disposition toward altruism and Christian love. To those who seemed to be predisposed toward altruism, Wesley likewise encouraged through regular meetings, since accountability had proven to be necessary to maintain changed behavior.

Wesley was not interested in working solely with people who already practiced altruism; in fact, he was rather inclined to help those on the fringes move to a lifestyle of habitual altruism. He sought often to reach those individuals who needed the greatest development toward it—those who Wesley felt had to be drawn out of spiritual ambiguity and into saving and changing grace. Wesley expressed clearly that he was not simply reorganizing those who were already Christians; to this end, he noted that he did not poach parishioners from the Anglican Church, claiming that those in his societies were mostly composed of "barefaced

---

[56]Snyder, *The Radical Wesley & Patterns for Church Renewal*, 57.
[57]Chilcote, *Recapturing the Wesleys' Vision*, 45.

heathens."[58] It is clear that Wesley saw the potential for people to shift away from their former lives into a new kind of society with a new kind of practice. Facilitating this shift required an intensive regimen, as historian Wesley Bebb states: "The Methodist church discipline of the eighteenth century has no parallel in modern English ecclesiastical history. [It] would be regarded as intolerable by almost all members of any Christian communion in this country to-day."[59] This statement seems a bit hyperbolic, because one can document contemporary movements with such rigor (monastic, new monastic etc.), but its sentiment rings true. In this way, the regiment of the world of constraints that Wesley created caused a group of people to shift on the selfish/selfless spectrum.

Looking at the human condition, John Wesley did not believe that humans were impossibly bent toward hedonism. After all, his concept of prevenient grace describes how God is continuously working within the lives of those who do not yet know him.[60] As Mildred Wynkoop observes, "Love, or holiness as [Wesley] interpreted it, was not the end of wholesome, even intense, human reactions but rather the disciplining of them."[61] Wesley capitalized on the energy generated by the spiritual renewal movement and disciplined the reactions to God's grace such that members could express their Godward and outward love as obedience to God's commands.

That this kind of discipline was not an easy process is evidenced in the fact that Wesley did not shy away from punitive practices when individuals needed to be rebuked: very regularly, people were excommunicated from the societies, classes and bands—in 1748, Wesley's Bristol society went from nine hundred down to 730.[62] In Wesley's journal, dated Saturday, March 12, 1743, Wesley also expelled, at one time, twenty-nine individuals for "lightness and carelessness."[63] However, the significance lies in the *cause* of expulsion. For example, it was rare that indi-

---

[58]Collins, *The Scripture Way of Salvation*, 220n85.

[59]Bebb, *Wesley: A Man with a Concern*, 123.

[60]Later in this chapter, I will show how Wesley's theological underpinnings helped him firmly establish the role of the bands and classes. His whole accountability structure has a reasoned theological foundation.

[61]Wynkoop, *A Theology of Love*, 17.

[62]Snyder, *The Radical Wesley & Patterns for Church Renewal*, 57.

[63]Wesley, *The Works of John Wesley*, 1:416.

viduals were expelled for strictly religious faults (not keeping with some ordinance of the Church of England), and none were ever recorded excommunicated for doctrinal differences, while the largest number were removed from fellowship for "not taking seriously enough their religion," and, according to Bebb, "to take it seriously always involved, in Wesley's view, right conduct to one's neighbour."[64]

Again, this harsh regiment was the connection between inward holiness and outward holiness that was necessary to keep the early Methodists accountable. In fact, Wesley focused so much on the connection between inward and outward holiness that he has been criticized as a legalist. There may have been a tendency for his early bands to slip into legalism, certainly a careful line that they had to walk. Wesley quickly repudiated the groups whose approach to problems was forensic, where holiness was legalistic and judgmental, instead of irenic, operating out of the love for God and neighbor.[65]

Wesley kept in check behavior connected to this outward holiness through various means, giving an account of the importance of connecting the heart to one's life in his journal entry in March 1747:

> Where is the difficulty then of finding out if there be any disorderly walker in this class, and, consequently, in any other? The question is not concerning the heart, but the life. And the general tenor of this, I do not say cannot be known, but cannot be hid without a miracle. . . . The society, which the first year consisted of above eight hundred members, is now reduced to four hundred. But, according to the old proverb, the half is more than the whole. We shall not be ashamed of any of these, when we speak with our enemies in the gate.[66]

It is not that Wesley took pride in losing numbers of his fledgling group; rather, he saw the loss of people as a pruning for proper growth wherein inward holiness reflected outward actions. Even within the groups one can see the emphasis on both the heart and the hands. Classes would meet one evening a week for an hour, during which time each person was compelled to divulge her or his particular needs and problems, report on spiritual progress and get the support and prayer from others in the

---

[64]Bebb, *Wesley: A Man with a Concern*, 128-29.
[65]Moore, "Classical Wesleyanism," 2.
[66]Wesley, *The Works of John Wesley*, 2:48-49.

group.[67] After the meeting, those involved experienced several hours of accountability where "advice or reproof was given as need required"[68] in order to encourage habitual behavior that would lead toward holiness. Interestingly enough, many people stayed in these highly regimented and personal groups, which was partly due to the fact that their leader was not demanding more than he was willing to do himself.

Wesley's conversion had much to do with the constraints he put on others. Convicted by two works that focus on total devotion to Christ, Wesley states in *A Plain Account of Christian Perfection* that "Mr. Law's 'Christian Perfection' and 'Serious Call' were put into my hands. These convinced me, more than ever, of the absolute impossibility of being half a Christian; and I determined, through his grace, (the absolute necessity of which I was deeply sensible of) to be all-devoted to God, to give him all my soul, my body, and my substance."[69] Thus, because the practice of his faith could not be defined by lukewarmness, Wesley was a complicated and regimented person. Such a mindset contributed to his fascination with the accountability and rigid routine he placed on his early bands and classes.[70] We find the fruits of his highly structured lifestyle in the great things Wesley accomplished in the course of his eighty-seven years: he rode 250,000 miles on horseback; preached over 45,000 sermons; published 233 original works on a myriad of subjects; compiled a Christian library; wrote a four-volume history of England; wrote a book called *Birds, Beasts, and Insects*; set up a free medical dispensary; adapted an electrical device for healing and allegedly cured more than one thousand people; set up multiple spinning and knitting shops for the poor; and received forty thousand

---

[67]Snyder, *The Radical Wesley & Patterns for Church Renewal*, 55.
[68]Wesley, *The Works of John Wesley*, 8:253-54.
[69]Ibid., 11:367.
[70]See Snyder, *The Radical Wesley & Patterns for Church Renewal*, 7. Snyder describes Wesley as a learned man who disciplined himself enough to achieve extraordinary things: "[Wesley] doesn't fit the molds in which we place him. We are not used to a popular mass evangelist who is also a university scholar, speaks several languages, knows classical and Christian authors by heart, and publishes his own English dictionary. Nor are we any better prepared to handle an evangelist who is also a social reformer or a theologian who preaches several times a day, develops his own discipling and nurturing system, sends out teams of traveling preachers, and publishes a home medical handbook that goes through twenty-some editions."

pounds[71] for his books and gave it all away.[72] Consequently, when Wesley demanded of his followers a high degree of regimen and discipline, it was not something foreign to him. This personal discipline helped to motivate his followers to press on even in the midst of structured constraints that were geared toward establishing inward and outward holiness.

Besides living this way himself, Wesley also required his followers to take their free actions and shoulder the responsibility, noting that "I am persuaded that every child of God has had, at some time, 'life and death set before him,' eternal life and eternal death, and has himself the casting voice. So true is that well known saying of St. Augustine . . . 'He that made us without ourselves, will not save us without ourselves.'"[73] For Wesley, God always treats individuals as responsible beings—his love does not override the integrity of another person,[74] and neither did Wesley's. Yet even within this individual freedom there was a highly constrained environment set up to encourage a certain kind of behavior. In fact, as Snyder mentions,

> Wesley did not permit discipline to grow lax. In his periodic visits to the
> various places he "examined," "regulated" or "purged" the classes and societies
> as need required. He (or later his assistants) would carefully explain the rules
> and exclude any who were not seeking to follow them. Excluded members
> would then receive no quarterly membership tickets. Many of these would
> later be readmitted if they mended their ways.[75]

These "tickets" were used as an entrance voucher to the celebratory love feasts that were regularly held. A member of a particular class would have to earn the right of admittance. Wesley would issue tickets to each class member that would have the individual's name, date and signature of Wesley or one of the leaders. This was the "proof" of their attendance.[76]

---

[71]This figure should be put in perspective with the average income of the late eighteenth century. A good yearly wage was merely a few hundred pounds. For more on this, see Kirstin Olsen, *Daily Life in 18th-Century England*, Daily Life Through History (Westport, CT: Greenwood Press, 1999), 140-45.

[72]Wynkoop, *A Theology of Love*, 61.

[73]Wesley, *Sermons 34-70*, 489-90.

[74]Moore, "Classical Wesleyanism," 17.

[75]Snyder, *The Radical Wesley & Patterns for Church Renewal*, 57.

[76]Ibid. See also Abel Stevens, *The History of the Religious Movement of the Eighteenth Century:*

The bands themselves had a rigorous set of constraints placed on the members by Wesley.[77] The following are some specific directions—according to the "Rules of the Band-Societies"—to those in bands concerning the poor: "Zealously to maintain good works; in particular,— 1) To give alms of such things as you possess, and that to the uttermost of your power. 2) To reprove all that sin in your sight, and that in love and meekness of wisdom. 3) To be patterns of diligence and frugality, of self-denial, and taking up the cross daily."[78] These questions all pertained to outward holiness and kept the members' actions in check. Yet these constraints were predicated on the discipline of inward holiness. Some questions (nonrhetorical) that were also asked of bands pertaining to inward holiness were the following: "1) What known sins have you committed since our last meeting? 2) What temptations have you met with? 3) How were you delivered? 4) What have you thought, said, or done, of which you doubt whether it be sin or not?"[79] Wesley had the expectation that people would progress in both their faith and actions. There was no room for static inward or outward holiness within the early Methodist bands.

The product of this combination of inward and outward holiness was plentiful. Wesley had his Methodists doing a whole host of ministries. These included the following:[80] setting up schools for children (including a grammar text Wesley himself wrote[81]), sick ministries, medical care, food and clothing distribution, ministry to unwed and destitute mothers, The Stranger's Friend Society (a charity for non-Methodists), ministry to paupers in London, establishing a home for widows in London, establishing an orphanage in Newcastle, unemployment relief, small business loan fund, and prison ministries.[82]

Called Methodism, Considered in Its Different Denominational Forms, and Its Relation to British and American Protestantism (New York: Carlton & Porter, 1859), 454-55. Evidently there were thirty-eight different types of tickets used between 1742 and 1765.

[77]See appendix three for a full set of "rules" and "directions" for the bands. Taken from Wesley, *The Methodist Societies*, 77-79.

[78]Wesley, *The Works of John Wesley*, 8:274. See also appendix three.

[79]Ibid., 273.

[80]See Stan Ingersol and Wesley Tracy, *Here We Stand: Where Nazarenes Fit in the Religious Marketplace* (Kansas City, MO: Beacon Hill Press, 1999), 10.

[81]See the "Grammar" section of Wesley, *The Works of John Wesley*, 14.

[82]For this list, I am also indebted to Moore, "Classical Wesleyanism," 6.

Wesley constructed a world of constraints to his highly organized young group. This resulted in sustained accountability and ordered behavior. Wesley introduced numerous measures to ensure that every member of his classes and bands were carefully looked after. Consequently, the "barefaced heathens" (Wesley's reference for the nonreligious) that composed the early Methodists tended to shift on the selfish/selfless spectrum, overcome biological constraints, adopt altruistic tendencies and mitigate egoism.

6.2 HOW WESLEY UNDERSTOOD AND NURTURED ALTRUISM BY WAY OF HOLINESS

John Wesley's understanding of human nature enveloped by prevenient grace influenced his understanding of altruism. He thought that inward holiness would lead to outward holiness, thus creating manifestations of practical altruistic living. One can see this in his sermon 91, "On Charity." He says,

> The sum of all that has been observed is this: Whatever I speak, whatever I know, whatever I believe, whatever I do, whatever I suffer; if I have not the faith that worketh by love, that produces love to God and all mankind, I am not in the narrow way which leadeth to life, but in the broad road that leadeth to destruction. In other words: Whatever eloquence I have; whatever natural or supernatural knowledge; whatever faith I have received from God; whatever works I do, whether of piety or mercy; whatever sufferings I undergo for conscience' sake, even though I resist unto blood: All these things put together, however applauded of men, will avail nothing before God, unless I am meek and lowly in heart, and can say in all things, "Not as I will, but as thou wilt!" We conclude from the whole, (and it can never be too much inculcated, because all the world votes on the other side,) that true religion, in the very essence of it, is nothing short of holy tempers.[83]

It is these "holy tempers" that Wesley sought to engender. Before looking deeper into the world of constraints he put on his followers, it is necessary to understand how Wesley understood the purpose and nature of altruistic good works.

---

[83]John Wesley, *Sermons 71–114*, ed. Albert C. Outler, The Bicentennial Edition of the Works of John Wesley (Nashville: Abingdon Press, 1986), 306.

Wesley understood altruism within the concept of the "social-love" of outward holiness. In his letter to Dr. Conyers Middleton, in "A Plain Account of Genuine Christianity," Wesley writes,

> His love to these, so to all mankind, is in itself generous and disinterested, springing from no view of advantage to himself, from no regard to profit or praise; no, nor even the pleasure of loving. This is the daughter, not the parent, of his affection. By experience he knows that *social love* (if it mean the love of our neighbor) is absolutely, essentially different from *self-love*, even of the most allowable kind, just as different as the objects at which they point. Yet it is sure that, if they are under due regulations, each will give additional force to the other, 'till they mix together never to be divided.[84]

Altruism, then, had to be understood within the context of community. Wesley's concept of the Christian life was dependent on a marriage between a personal encounter with Christ and a shared experience within a close Christian community.[85] Concerning such, Wesley states that to not have this kind of self-giving love toward others would be to err in committing spiritual adultery. Accordingly, the social love is what helps draw one toward Christian perfection. He says,

> "Holy solitaries" is a phrase no more consistent with the gospel than holy adulterers. The gospel of Christ knows of no religion, but social; no holiness but social[86] holiness. "Faith working by love" is the length and breadth and depth and height of Christian perfection. "This commandment have we from Christ, that he who loves God, love his brother also"; and that we manifest our love "by doing good unto all men; especially to them that are of the family of faith."[87]

---

[84]Wesley, *John Wesley: A Representative Collection of His Writings*, ed. Albert C. Outler (New York: Oxford University Press, 1964), 184-85.

[85]Chilcote, *Recapturing the Wesleys' Vision*, 48.

[86]It is important to note that the word *social* does not mean "social justice" but rather "communal."

[87]John Wesley and Charles Wesley, *Hymns and Sacred Poems*, 3rd ed. (London: W. Strahan, 1739), 5. There are numerous examples of language like this. Take for instance Wesley's words in his sermon 74, "Of the Church," where he says, "In the meantime, let all those who are real members of the Church see that they walk holy and unblameable in all things. . . . Show them your faith by your works. Let them see, by the whole tenor of your conversation, that your hope is laid up above! Let all your words and actions evidence the spirit whereby you are animated! Above all things, let your love abound; let it extend to every child of man; let it overflow to every child of God. By this let all men know whose disciples ye are, because you love one another." See Wesley, *Sermons 71-114*, 56-57.

These are Wesley's words shortly after his conversion experience, in the midst of the early 1739 revival movement.

For Wesley, being an altruistic community was not an exaggerated dream. He thought that it was not only absolutely possible but also absolutely necessary. In his sermon 122, "Causes of the Inefficacy of Christians," Wesley tellingly answers the rhetorical question,

> "But is it possible to supply all the poor in our society with the necessaries of life?" It *was* possible once to do this, in a larger society than this. In the first Church at Jerusalem "there was not any among them that lacked; but distribution was made to every one according as he had need." And we have full proof that it may be so still. It is so among the people called Quakers. Yea, and among the Moravians, so called. And why should it not be so with *us*? "Because they are ten times richer than us." Perhaps fifty times. And yet we are able enough, if we were equally willing, to do this. A gentleman (a Methodist) told me some years since, "I shall leave forty thousand pounds among my children." Now suppose he had left them but twenty thousand, and given the other twenty thousand to God and the poor; would God have said to him, "Thou fool?" And this would have set all the society far above want.[88]

There are numerous anecdotes that show the zeal of the early Methodists.[89] The enthusiasm is often centered on the outward expression of

---

[88]John Wesley, *Sermons 115–151*, ed. Albert C. Outler, The Bicentennial Edition of the Works of John Wesley (Nashville: Abingdon Press, 1987), 92.

[89]One conversation of a Methodist minister and a new convert: "'What are your rules?' inquired Elias. 'One penny a-week, and one shilling a-quarter,' replied the minister, 'were the contributions which Mr. Wesley required of every member, unless in extreme poverty.' Father Damper glanced curiously at Elias, evidently thinking that now, at least, a test had come to prove the sincerity of the new convert. 'A penny a-week and a shilling a-quarter,' said Elias, 'to support the ministry and the work of God. I'm afraid the work of God will not be very strong if its best friends support it with only a penny a-week and a shilling a-quarter. That's twopence a week. How many twopences have I spent in sin! How many twopences a week have I wasted in ways of shame, to ruin both body and soul for ever! And now that Divine grace has done so much for me—the light of truth in my mind, the peace of God in my conscience, the love of God in my heart, the blessing of God in my home, a clear head for my business, and a better character to help it—have I nought more to support the ministry and the work of God than a penny a-week and a shilling a-quarter? I could not sleep on't. I could not take so much and give so little. Have not I been singing this very night—*Too much to Thee I cannot give, Too much I cannot do for Thee; Let all Thy love, and all Thy grief, Graven on my heart for ever be.* How would a penny a-week and a shilling a-quarter look in the light of such a seal as that? My Saviour giving to me the "pearl of great price," and I giving back to Him *two bits of brass*—only the price of a pot of beer at Sam Spigot's! Nay, nay; I should like to start fairer with the Master, and fairer with ye, than that. Put me down sixpence a-week and ten shillings a-quarter. That's little enough. Many a time have I

holiness. Monies that were raised almost always went to aid the poor. The "penny a week and a shilling a quarter" that was the norm raised considerable sums that were later used for the underprivileged.[90] It is in this vein that Wesley says one should "give all you can."[91]

Besides nurturing altruism by way of holiness, John Wesley (whether purposefully or not) used holiness to mitigate egoism. Wesley states in *A Plain Account of the People Called Methodists*, "We introduce Christian fellowship where it was utterly destroyed. And the fruits of it have been peace, joy, love, and zeal for every good word and work."[92] This introduction of a new kind of Christian fellowship was for the purpose of calling back the church to holiness. These actions ultimately reined back the egoistic tendencies of its members. Individuals who are justified by faith are moved by the Holy Spirit to uncover hidden sins and to recognize temptations that, as Outler explains, "If entertained seriously enough to form moral intentions, will result in the forfeiture of one's justification."[93] Thus, one main focus for Wesley's early groups was to be called to a life of tempered action.

True to form, Wesley connected the mitigation of egoism to an issue of outward holiness that stemmed from one's inward holiness. In sermon

---

spent more in one night in support of Sam Spigot and the "Black Bull."" See John Bate, ed., *The Local Preacher's Treasury* (London: Lile and Fawcett, 1884), 149.

[90]Snyder, *The Radical Wesley & Patterns for Church Renewal*, 55.

[91]The exact quotation comes from his sermon 50, "The Use of Money." Here he suggests a fuller understanding of this to earn, save and give all one can. Wesley states, "But let not any man imagine that he has done anything, barely by going thus far, by 'gaining and saving all he can,' if he were to stop here. All this is nothing, if a man go not forward, if he does not point all this at a farther end. Nor, indeed, can a man properly be said to save anything, if he only lays it up. You may as well throw your money into the sea, as bury it in the earth. And you may as well bury it in the earth, as in your chest, or in the Bank of England. Not to use, is effectually to throw it away. If, therefore, you would indeed 'make yourselves friends of the mammon of unrighteousness,' add the Third rule to the two preceding. Having, First, gained all you can, and, Secondly saved all you can, Then 'give all you can.'" See Wesley, *Sermons 34–70*, 276-77. A similar statement by Wesley employs the same rhetoric, found in sermon 122, "Causes of the Inefficacy of Christianity": "Who regards those solemn words, 'Lay not up for yourselves treasures upon earth?' Of the three rules which are laid down on this head, in the sermon on 'The Mammon of Unrighteousness,' you may find many that observe the First rule, namely, 'Gain all you can.' You may find a few that observe the Second, 'Save all you can:' But how many have you found that observe the Third rule, 'Give all you can?' Have you reason to believe, that five hundred of these are to be found among fifty thousand Methodists? And yet nothing can be more plain, than that all who observe the two first rules without the third, will be twofold more the children of hell than ever they were before." See Wesley, *Sermons 115–151*, 91.

[92]Wesley, *The Works of John Wesley*, 8:251-52.

[93]Albert C. Outler, *Theology in the Wesleyan Spirit* (Nashville: Tidings, 1975), 39.

123, "The Deceitfulness of the Human Heart," Wesley mentions how
these issues of egoism are derived from spiritual issues. He says,

> Only let it be remembered, that the heart, even of a believer, is not wholly
> purified when he is justified. Sin is then overcome, but it is not rooted out; it
> is conquered, but not destroyed. Experience shows him, First, that the roots
> of sin, self-will, pride, and idolatry, remain still in his heart. But as long as he
> continues to watch and pray, none of them can prevail against him. Expe-
> rience teaches him, Secondly, that sin (generally pride or self-will) cleaves to
> his best actions: So that, even with regard to these, he finds an absolute ne-
> cessity for the blood of atonement.[94]

In this way, Wesley was trying to warn people of their spiritual action, as
he saw their egoistic works as a life of bondage. As Outler says, "The fruit
of sin is *bondage* (i.e., slavery to our own self-deceptions, to our illusions
about life and society that stir up utopias that never quite transpire)."[95]
Wesley's admonition against egoism is also predicated on warning people
against practical aspects of holiness. In sermon 87, "The Dangers of
Riches," Wesley says,

> I ask, then, in the name of God, Who of you "desire to be rich?" Which of *you*
> (ask your own hearts in the sight of God) seriously and deliberately desire
> (and perhaps applaud yourselves for so doing, as no small instance of your
> *prudence*) to have more than food to eat, and raiment to put on, and a house
> to cover you? Who of you desires to have more than the plain necessaries and
> conveniences of life? Stop! Consider! What are you doing? Evil is before you!
> Will you rush upon the point of a sword? By the grace of God, turn and live![96]

It is precisely this kind of zeal that inspired many Methodists to examine
and mortify their varied bondages to sin and to consider others before
themselves.

6.3 ENVIRONMENTAL CONSTRAINTS THAT TEMPER BIOLOGICAL CONSTRAINTS

As we might recall from previous chapters, environmental constraints
have the potential to temper biological constraints placed on humans

---

[94]Wesley, *The Works of John Wesley*, 7:341.
[95]Outler, *Theology in the Wesleyan Spirit*, 40.
[96]Wesley, *Sermons 71–114*, 236. Emphasis his.

from birth. Each human being is born with her or his own set of genetic influences that modify and guide behavior, such as their agility or propensity toward alcoholism. Yet, Wesley structures his early Methodist groups to lead regimented lives full of accountability that necessarily mitigate certain biological influences that might incline one toward egoism and instead feed one's inclinations toward altruism (a constituent part of holiness). Humans, of course, do not live in isolation—community has always been part of the context in which humans live. So it is no coincidence that although evolutionary theory claims that it is difficult to learn how to become altruistic individuals, evolved traits are always trained within the framework of community.[97] Consequently, each community has the potential to push the boundaries of biological constraints wider and wider.

When discussing this kind of community in which individuals might be encouraged to live more altruistically (or even more ethically), Stephen Pope, in his *Human Evolution and Christian Ethics*, discusses how sociobiologist and philosopher Michael Ruse claims that when it comes to Christian ethical behavior, Christians "talk a good game," especially about love. But most Christians live lives that bear little resemblance to the high ethic of the New Testament.[98] While this might be anecdotally true, it is not wholly so. In response to this, Pope makes a remarkably Wesleyan statement concerning the struggle for those living within Christian community (like ones established by Wesley) to perform acts of altruism. Pope's statement reflects the concept of prevenient grace. He lists four ways to interpret the reason why individuals may struggle to display as much love as they would like to:[99] (1) Any theory that is worth anything is difficult to hold up in practice; (2) a Lutheran response is that evolutionists misunderstand that moral law was given only to condemn sin and show we are all utterly dependent on God's grace; (3) Augustinian models show that Christian community looks like the world because God allows free will (good and bad) to live together and

---

[97]Take for example, human sexuality. Though this is a natural trait, it is conditioned by a host of "norms" from specific communities. See Stephen J. Pope, *Human Evolution and Christian Ethics*, New Studies in Christian Ethics (Cambridge: Cambridge University Press, 2007), 226-27.

[98]Ibid., 233.

[99]For reference to the following list, please see ibid., 233-34.

then judges at the end of time. The city of God is hidden and God alone knows its members; and (4) God alone is capable of judgment on the human heart. But since God desires the salvation of the entire world, grace is silently and subtly operative everywhere and in every human life (prevenient grace). People respond to this offer of grace in ways made available by their particular cultural and historical contexts. Pope further states,

> Because of the complexity of history, Christians have been both slave owners and abolitionists, soldiers and pacifists, pro-choice and pro-life. Christians are well prepared to acknowledge the historical limitations imposed on their own judgments. This is precisely the major weakness of the evolutionists—the failure adequately to acknowledge the historical and cultural nature of the human being. Whatever universal species-specific biological traits we have will always bear their moral significance within particular cultural contexts.[100]

Evolution puts us at a place where humans can still be influenced by the environment in which they dwell. Yet dwelling in a certain kind of community might be necessary for Christians to be more altruistic in practice. And Wesley shows us what that might look like. Christian ethics can offer helpful counsel to sociobiologists by stating that we should not confuse our natural propensity toward self-love with the full range of our moral responsibilities to others.[101]

6.4 CONCLUSION

In this chapter I hoped to show that as human nature leaves the possibility for one to shift on the egoism/altruism spectrum (discussed in previous chapters), Wesley took this to his advantage to promote the holistic altruism that was at the heart of his drive toward holiness. When individuals dwelled within these purposeful communities, they exhibited the distinguishing virtue of altruism. One advantage of living in the era of de Waal is that we can understand that the human condition is not totally depraved egoism, and by following Wesley's example can take Christian generosity to even higher levels. John Wesley's small

---

[100]Ibid., 234.
[101]Ibid., 215.

groups did this by capitalizing on a natural phenomenon by creating a world of constraints full of accountability.

There is also no doubt that John Wesley was a catalyst for spiritual revival across England (and into America). Moreover, he renewed what it meant to be a Christian living in the world through tangible and practical service to social needs. No one can say of Wesley that he was "so heavenly bound he was no earthly good." Any suggestion like this would, as J. Wesley Bready says, provide "too little veracity for even the sorriest caricature" of who Wesley is.[102] For Wesley, giving and having to live among the poor was part of working out one's salvation. In regards to this, in sermon 122, "Causes of the Inefficacy of Christians," Wesley states,

> But I will not talk of giving to God, or leaving, half your fortune. You might think this to be too high a price for heaven. I will come to lower terms. Are there not a few among you that could give a hundred pounds, perhaps some that could give a thousand, and yet leave your children as much as would help them to work out their own salvation? With two thousand pounds, and not much less, we could supply the present wants of all our poor, and put them in a way of supplying their own wants for the time to come.[103]

In this way, and countless others, Wesley encourages his followers to live alongside the poor. Being socially active was a key component for John Wesley's ministry and discipleship model.[104] To be sure, Wesley lived out what he preached. Again, in "Causes of the Inefficacy of Christians," Wesley states,

> Do you gain all you can, and save all you can? Then you must, in the nature of things, grow rich. Then if you have any desire to escape the damnation of

---

[102]Bready, *England*, 225.

[103]Wesley, *Sermons 115–151*, 92-93. Some other phenomenal pieces from that sermon include Wesley chiding his members for not practicing such. He says, "But to return to the main question. Why has Christianity done so little good, even among us? Among the Methodists,— among them that hear and receive the whole Christian doctrine, and that have Christian discipline added thereto, in the most essential parts of it? Plainly, because we have forgot, or at least not duly attended to, those solemn words of our Lord, 'If any man will come after me, let him deny himself, and take up his cross daily, and follow me.' It was the remark of a holy man, several years ago, 'Never was there before a people in the Christian Church, who had so much of the power of God among them, with so little self-denial.' Indeed the work of God does go on, and in a surprising manner, notwithstanding this capital defect; but it cannot go on in the same degree as it otherwise would; neither can the word of God have its full effect, unless the hearers of it 'deny themselves, and take up their cross daily.'" See ibid., 93-94.

[104]Chilcote, *Recapturing the Wesleys' Vision*, 50.

hell, *give* all you can; otherwise I can have no more hope of your salvation, than of that of Judas Iscariot. I call God to record upon my soul, that I advise no more than I practise. I do, blessed be God, gain, and save, and give all I can. And so, I trust in God, I shall do, while the breath of God is in my nostrils. But what then? I count all things but loss for the excellency of the knowledge of Jesus my Lord! Still, *I give up every plea beside, "Lord, I am damn'd—but thou hast died!"*[105]

It is in John Wesley that we find a self-conscious individual who intuitively understood the human condition to be moveable on the selfish/selfless spectrum. With the creation of his world of constraints, Wesley's intention was to move people toward inward holiness that resulted in outward holiness. This environment helped nudge Wesley's followers ever closer to what it means to live a life sacrificially for others.

---

[105]Wesley, *Sermons 115–151*, 96. The last italicized sentence comes from John Wesley and Charles Wesley, *Hymns and Sacred Poems*, 94.

7

A LIFESTYLE OF HOLINESS

◆

Despite recent advances in the field of sociobiology, theological ethics is still coming to terms with what this new knowledge means for how we understand moral behavior. It has been my attempt, then, to locate this book at the intersection of theological ethics and the sciences and begin to answer some of the following questions: How much does Wesleyan ethics connect to sociobiology, and can these two fields fully explain moral traits such as human altruism? If genetic explanations do not fully explain human altruism, what role should we give to environmental explanations and free will? How do genetic explanations of altruism relate to theological accounts of human goodness? In order to work toward answers to these questions, I have used the lens of Wesleyan ethics, hoping through its application to have argued how such a perspective offers a fresh assessment within the interface of sociobiology and ethics. What is more, reading these questions through the perspective of Wesleyan ethics has brought me to the following conclusions: Wesleyan bands and classes can provide the environmental conditions within which people may develop, beyond their genetic inclinations, holiness. These groups were a product of John Wesley's theological understanding of Christian perfection against a background of evolutionary biology (of which he was intuitively aware). Consequently, it has also been a chief aim of this book to explore the significant role sociobiology plays in our understanding of Wesleyan ethics.

As I looked at how John Wesley encouraged an environment by which an individual could more easily move toward holiness, it was altruistic action[1]—or what Wesley often called "social" holiness or "outward" holiness—that was often at the heart of those communities. Thus, just as sociobiologists would not consider using language such as "holiness," Wesley was unfamiliar with the field of sociobiology or even the term *altruism* (coined in the nineteenth century). Instead of using the language of holiness, sociobiologists tend to talk about altruism; Wesley considered altruistic practices to be a result of holiness. Altruism, however, is not without a complicated history within the field of evolutionary biology. Until recently, it was thought that human genetics had predisposed humans to egoism, causing altruism to become a puzzling phenomenon within sociobiology, creating problems for those who solely appeal to genetic explanations. Yet even within the field of biology, recent studies show that the biological composition of individuals puts them more to the center of the selfish/selfless spectrum, not solely selfish as once thought.[2] Therefore, despite the confusion within sociobiology and the difference in terminology between holiness and altruism, there are significant connections to be made between the two concepts. It is

---

[1]As noted earlier in the book—though it is worth repeating here—I realize that the word *altruism* comes with much debate. Some, like Neil Messer, argue that altruism is an anemic word that is enigmatic within Christian circles. I agree with Messer that altruism does not straightforwardly name a "virtue" within Christian ethics and is often employed sloppily. Yet, in the way sociobiologists use the word, I do not feel that it is altogether different. Consequently, I decided to use this phrase throughout my book. In regard to defining altruism, Stephen Pope offers an interesting distinction between morally good actions and altruistic ones, claiming, and I think rightly so, that not all altruism is morally justified (he cited terrorism and how it can be considered a form of altruism). But for the purposes of this book, I used the word *altruism* to describe actions that are directly linked to the well-being of others. See Stephen J. Pope, *Human Evolution and Christian Ethics*, New Studies in Christian Ethics (Cambridge: Cambridge University Press, 2007), 227-28. Furthermore, there is a distinction between altruism as intention, and altruism as behavioral consequence that benefits other organisms without corresponding benefits to oneself, which can take place without any conscious moral reasoning or choice. For more on this, see Sarah Coakley, *Sacrifice Regained: Evolution, Cooperation, and God*, Gifford Lectures (University of Aberdeen, 2012), www.abdn.ac.uk/gifford/about/2012-giff/.

[2]As noted earlier in this book, despite the changing thinking within the field of evolutionary biology, evolutionary theorists acknowledge that there has been little major genetic evolution in the human species over the last fifty thousand years. This is true at least in the sense that there has been no human speciation in this interval, and that cultural change has had the bulk of the influence on human behavior. See Jared M. Diamond, *The Third Chimpanzee: The Evolution and Future of the Human Animal* (New York: HarperCollins, 1992). See also Pope, *Human Evolution and Christian Ethics*, 162. See also Kenneth M. Weiss and Anne V. Buchanan, *The Mermaid's Tale: Four Billion Years of Cooperation in the Making of Living Things* (Cambridge, MA: Harvard University Press, 2009).

thus into the fog of such initial confusion that John Wesley's ethics allows us to shed some important light.

Wesley intuitively[3] picked up on the natural phenomenon of the selfish/selfless spectrum, demonstrating that humans can indeed move along this spectrum when he created intentional communities that shifted humans who were genetically inclined toward selfishness to become holier—and, consequently, more altruistic. This type of intentional community was at the heart of the early Methodist movement. When individuals resided within Wesley's highly organized community, they learned traits and qualities that were characteristic of altruism, as can be observed in Wesley's bands and classes. These groups were the engines of the communal lifestyle of altruism toward the poor.[4] In this way, sociobiology confirms what Wesley knew to be true of human nature: that humans who dwell within a system of constraints motivated by Christian perfection cultivate altruistic behavior.

In this book I have sought to demonstrate how Wesleyan ethics could provide a framework for approaching sociobiology. I have also sought to establish the connection between the sociobiological definition of cooperative behavior—which rests somewhere between egoism and altruism—and accountability groups, which have the potential to engender generosity and altruism. What is more, I have articulated how John Wesley himself distinctively transmitted his social ethic through the organized bands that held members answerable to a high standard. In regard to the social context of his time, the small groups formed by Wesley (his bands and classes) encouraged people to live out a more generous social ethic through accountability. In the end, this project works toward a fuller account of how Wesleyan groups bridged the gap between genetic/environmental constraints and ideas of Christian perfection; it has thus been my attempt—though on the surface it may appear a contradiction—to articulate clearly how

---

[3]It would be anachronistic to say that Wesley knew anything about sociobiology or even evolution. Yet Wesley understood and recognized the human condition beyond what was common for a man of his day and age.

[4]The kind of intentional community is evaluated depending on how richly the moral character and activity is engendered. See Stanley Hauerwas, *A Community of Character: Toward a Constructive Christian Social Ethic* (Notre Dame, IN: University of Notre Dame Press, 1981), 34.

these concepts can form a cohesive and holistic account of the Christian person.[5]

7.1 BRIEF SUMMARY OF MAIN CHAPTERS

In this book I began with a sociobiological look at altruism, including a critique of the inability of sociobiology to fully explain altruism, and an introduction to both free will and environmental constraints on that free will, concluding that Wesleyan ethics might express its own account of altruistic behavior as a theological concept of Christian perfection—a concept that drove John Wesley to develop his own set of environmental constraints on behavior to help encourage people to be more altruistic.

It was my goal in chapter two to help the reader understand basic and current sociobiological explanations of altruistic behavior. Here, I discussed basic Darwinian history and how altruism has been a stumbling block for evolutionary theorists who have attempted, unsuccessfully, to explain why or how it exists. To address the presence of altruism among humans, I elucidated kin selection theory, group selection theory (also called multilevel selection theory) and game theory, seeking to provide clarification of current research within the field of sociobiology.[6]

---

[5]Mark H. Mann has an interesting perspective in *Perfecting Grace* that speaks to this "holistic" person. He says, "If our answers to the questions that people bring with their quest for the holy life are overly simplistic and without sensitivity to the complexity and particularity of their lives, then our answers can have devastating effects. If such persons are provided with cookie-cutter answers that have little to do with the true complexity of their lives, then we leave them subject to one of two alternatives: Either they must deny the complexity of their lives in order to make themselves fit the mold prescribed for them, or they must conclude that their lives do not fit the mold. In either case, these persons are sold short of the fullness of holiness." See *Perfecting Grace: Holiness, Human Being, and the Sciences* (New York: Bloomsbury Academic, 2006), 175.

[6]Stephen Pope has a wonderful statement that is too large to quote in text, but speaks to what I was hoping to accomplish in this chapter, namely setting up the "altruism problem." He says, "Evolution provides an account of why we are emotionally inclined to learn to care for self, our own families, friends, and communities more than for others. Since evolved predispositions are always trained in the context of particular culture and communities, the specific shapes they take in concrete lives show significant variation. Evolutionists insist that it is hard to learn genuine altruism, that is, altruism from non-kin and nonreciprocators. An inclination for genuine altruism cannot have evolved because it would not have been 'selected for': over the span of evolutionary time, organisms that practiced genotypical altruism would have been less likely to reproduce (less 'inclusively fit') than their genotypically egotistic counterparts, so they would have been reproductively eliminated from their respective populations. The ultimate genotypical altruists—the parent who feeds other people's children while her children go hungry—would not have left as many descendants as the genotypical egoists. This claim amounts to the uncontroversial statement that human nature cannot have evolved in such a way that individuals would consis-

Chapter three offered a critique of the sociobiological explanations of altruism, in which I discussed the storied relationship between sociobiology and altruism, examining the inability of sociobiology to adequately explain altruism. I provided arguments for the reductionist nature of the work[7] of prominent sociobiologists such as E. O. Wilson and Richard Dawkins, while introducing the individual who would be the focus of the rest of the book: John Wesley. The inability of sociobiology to provide a full accounting of altruism, when held up against religious revival movements like Wesley's,[8] finds sociobiology to posit unsatisfactory explanations concerning the holistic human.

Next, in chapter four, I discussed how the biological and environmental constraints on human behavior influence where individuals are located on the selfish/selfless spectrum. I argued that, although all humans are influenced by some internal and external constraints, they still have the unique capacity to freely move on that continuum.[9] This

---

tently prefer to promote the well-being of others to their own well-being, to prefer to promote the well-being of cheats to reciprocators, or to sacrifice the good of their own community for the good of other communities." Pope, *Human Evolution and Christian Ethics*, 226-27.

[7]As noted in chapter three, these reductionist tendencies often border on suprascience. As Henry Plotkin says, "Underlying all the biological and social sciences, the reasons for it all, is the 'need' (how else to express it, perhaps 'drive' would be better) for genes to perpetuate themselves. This is a metaphysical claim, and the reductionism that it entails is . . . best labeled as metaphysical reductionism. Because it is metaphysical it is neither right nor wrong nor empirically testable. It is simply a statement of belief that genes count above all else." Henry C. Plotkin, *Evolution in Mind: An Introduction to Evolutionary Psychology* (Cambridge, MA: Harvard University Press, 1998), 94.

[8]It should be noted that when Wesley discussed the personal and social side of religious holiness, he most often spoke of "personal holiness" and "social virtue." Yet his language conveys a kind of faith that is impacted by a whole host of influences that move individuals simultaneously toward God and toward each other. See Paul Wesley Chilcote, *Recapturing the Wesleys' Vision: An Introduction to the Faith of John and Charles Wesley* (Downers Grove, IL: InterVarsity Press, 2004), 52.

[9]For Thomas Oden, freedom is also considered a grace from God. He writes, "The doctrine of grace is an argument for human freedom. It would be more absurd if God had worked in a costly way to free us, yet we remained automatons or puppets. God would not work in us to free us were we not created with the capacity for freedom, which though now fallen into sin, can be redeemed and reconstituted by grace." Thomas C. Oden, *John Wesley's Teachings: Christ and Salvation*, vol. 2 (Grand Rapids: Zondervan, 2012), 145. To reference this idea back to John Wesley, see the sermon "The Deceitfulness of the Human Heart," in John Wesley, *Sermons 115-151*, ed. Albert C. Outler, The Bicentennial Edition of the Works of John Wesley (Nashville: Abingdon Press, 1987), 150-60. Oden also writes, "As God creates us *ex nihilo* without any cooperation of our own, for no one makes an application to be born, so God recreates our freedom to love, rescuing us from our fallen condition of unresponsive spiritual deadness." Oden, *John Wesley's Teachings: Christ and Salvation*, 146.

chapter paved the way for a discussion of John Wesley's environmental constraints.

Chapter five focused on Wesley and the connection between the sociobiological understanding of altruism and Wesley's theological understanding of both original sin and Christian perfection. In order to fully account for Wesley's unique bands and classes (those groups that provided a system through which Wesley encouraged altruism by way of social holiness), it was necessary to work through his underlying theological presuppositions. Wesley's longest treatise, *The Doctrine of Original Sin*, highlights the inability of fallen humanity to act righteously apart from the divine grace of God[10]—a bleak notion of the human condition that was quite Augustinian in nature.[11] Yet, for Wesley, humans were never without the prevenient grace of God; and Wesley did not hesitate to hold these two concepts (total depravity and prevenient grace) in tension.[12] For instance, in John Wesley's *Remarks on Mr. Hill's Review*, he states that the "will of man is by nature free only to evil. Yet . . . every man has a measure of free-will restored to him by grace."[13] These

---

[10]See Oden, *John Wesley's Teachings: Christ and Salvation*, 150. As Oden says, "Calvin's doctrine of common grace forms the background of Wesley's teaching of preparatory grace that precedes saving grace. This teaching stood on the shoulders of Augustine, whose writings are saturated with a high doctrine of grace, just as we have seen in Wesley." See ibid., 149.

[11]Wesley's doctrine of grace is heavily influenced by, but not limited to, Augustine—seeing God's favor at work throughout Wesley's entire *ordo salutis*. For Augustine, this would be common grace, saving grace and completing grace. Thomas Oden mentions that there are not three separable graces but merely one "grace" working through the process. See ibid., 137. This would include a "total depravity" kind of foundation (apart from prevenient grace). Oden writes, "Assuming the depth of the drastic human predicament as spelled out in 'The Doctrine of Original Sin,' it is impossible without grace to make the least motion toward God." See ibid., 140. See also John Wesley, *Sermons 71–114*, ed. Albert C. Outler, The Bicentennial Edition of the Works of John Wesley (Nashville: Abingdon Press, 1986), 202-3. Also, as Oden suggests, "All phases of salvation are permeated by grace that we may be led both to will and to do. Willing and doing lead Wesley to make a distinction between inward and outward actions: inward religion (holiness of the heart) is grounded in God's work in us '*to will*' (*to telein*, 'to desire, wish, love, intend'). Outward religion (holiness of life) is grounded in God's giving us the energy '*to do*' (*to energein*, 'to energize, execute, actualize') his good pleasure. This energy that comes from God 'works in us every right disposition, and then furnishes us for every good word and work.'" See Oden, *John Wesley's Teachings: Christ and Salvation*, 140-41.

[12]This grace (both for prevenient grace that influences original sin and the grace that influences Christian perfection) works through gradual and instantaneous means. See Oden, *John Wesley's Teachings: Christ and Salvation*, 145. See also John Wesley, *The Letters of the Rev. John Wesley*, ed. John Telford, standard ed., 8 vols. (London: Epworth Press, 1931), 7:267; see also 2:280.

[13]John Wesley, *The Works of John Wesley*, ed. Thomas Jackson and Albert C. Outler, 14 vols. (Grand Rapids: Zondervan, 1958), 10:392. It would be against the teachings of Wesley to think

arguments on prevenient grace are not about natural abilities to be able to will or be free, but rather about grace working *through* nature.[14] As Oden articulates, "When we cooperate with the unmerited grace of God's saving act on the cross, we [should] not forget that it is precisely grace that *enables* our cooperation."[15]

In chapter six, I further unpacked John Wesley's ideas, illustrating how his understanding of holiness and Christian perfection led him to develop a complete system of checks and balances[16]—which I call his "world of constraints"—that created an environment whereby individuals were much more likely to move toward the altruistic side of the selfish/selfless spectrum. The idea of Christian perfection and its intersection with biological predispositions is not without complications. It should be clearly stated that in the Wesleyan perspective, no one does good work without grace.[17] So the good work that one does within Wesley's groups is a grace working with the human condition, not the human condition working autonomously and divorced from the grace of God. Therefore, I worked through these possible critiques and conveyed the idea that within a community concerned with Christian perfection, an individual can develop a virtuous character that nurtures the overcoming of genetic constraints.

7.2 HOLINESS OUTSIDE THE WESLEYAN COMMUNITY

Wesleyan notions and practices of holiness are both particular and

---

that grace was part of the natural, fallen human condition; instead, grace, as seen in Wesley's Sermon 85, "On Working Out Our Own Salvation," is a heavenly gift that is completely unmerited and foreign to the totally depraved human. Yet, for Wesley, it so happens that no human is without prevenient grace, which alters the original sin and totally depraved individual. Likewise, grace is given to those to attain perfection. See Wesley, *Sermons 71–114*, 202-3. See also Oden, *John Wesley's Teachings: Christ and Salvation*, 151.

[14]See Oden, *John Wesley's Teachings: Christ and Salvation*, 152.

[15]Ibid. Emphasis mine. Oden goes on to say, "Though not intrinsic to freedom, grace is constantly present to freedom as an enabling, wooing gift. That does not reduce grace to an expression of nature. Grace remains grace. It is not something we possess by nature. It is given us. Yet grace is given abundantly to everyone, from the Paleolithic mound makers of Georgia to the forest Hottentots of Africa. Everywhere human beings exercise freedom, there grace is working to elicit, out of the distortions of fallen human nature, responses of faith, hope, and love. Preparatory grace remains a teaching that can be twisted so as to imagine that Wesley was covertly affirming the very Pelagianism he so frequently denied." See also Wesley, *Sermons 71–114*.

[16]For Wesley, exegesis and living are one and the same. See Carl Bangs, *Our Roots of Belief: Bible Faith and Faithful Theology* (Kansas City, MO: Beacon Hill Press, 1981), 41.

[17]Oden, *John Wesley's Teachings: Christ and Salvation*, 141.

stringent. Consequently, I saw fit to use John Wesley as a lens to connect holiness to evolution as well as altruism to Christian living. If holiness can connect to evolution here, then it might be able to connect more easily to other Christian persuasions. Yet, while this book dives deeply into Wesleyan holiness, if the reader does not affiliate with this branch of Christianity, she or he can find similar principles that connect to another particular tradition. While it was not the task of this book to tackle various theologies of holiness outside the Wesleyan tradition, it was a secondary hope of mine that those who hold to differing understandings of holiness could extrapolate broader generalizations to their own faith. Below are some examples from a select group of Christian traditions that might have similar connections between holiness and evolution just as the Wesleyans.

In true form to John Wesley's ecumenical spirit, Wesley was influenced by both the Roman Catholic tradition (through Anglicanism) as well as the Eastern Orthodox tradition, as we have seen in his adaptation of *theosis*. Thus, the reader will find many similarities between them. I will refrain from recapping too much material here, but will instead point the reader to some works that connect holiness to evolution. As far as Catholicism is concerned, I have relied heavily on Stephen Pope, whose surname is probably not coincidental. Pope does an excellent job bringing together both evolutionary acts of altruism and love in his work, *The Evolution of Altruism and the Ordering of Love*.[18] Here, a very good Catholic understanding of both the practical displays of holiness as well as evolutionary influences on love and altruism are brought together.

On the other side of Christianity, as noted in previous chapters, the Eastern Orthodox Church has much in common with Wesleyan notions of holiness. With his desire to navigate a *via media* between Protestantism and the Catholic Church (as this is one of Anglicanism's aims), Wesley almost seems to stumble upon Eastern Orthodoxy.[19] As Randy Maddox states in one article, "Even a cursory reading of Wesley shows that these recovered early Greek theological voices were important to

---

[18]Stephen J. Pope, *The Evolution of Altruism and the Ordering of Love*, Moral Traditions & Moral Arguments (Washington, DC: Georgetown University Press, 1994).

[19]K. Steve McCormick, "Theosis in Chrysostom and Wesley: An Eastern Paradigm of Faith and Love," *Wesley Theological Journal* 26 (1991).

him."[20] Wesley frequently cited Basil, Chrysostom, Clement of Alexandria, Clement of Rome, Ephraem Syrus, Ignatius, Irenaeus, Justin Martyr, Origen, Polycarp and Macarius.[21] Besides *theosis*, Wesley definitely had much similarity with the Eastern fathers, especially when dealing with the full spectrum of Macarius's work.[22] While the controversy over whether Wesley was more Eastern than Western is still alive, much of this book's discussion of holiness has an Eastern leaning. If the reader might be interested in exploring this further, she or he might start by looking at Christoforos Stavropoulos's chapter in *Eastern Orthodox Theology* titled "Partakers of Divine Nature."[23]

When exploring holiness in other Protestant traditions besides Wesleyanism, the reader might be interested in the multiauthored book *Five Views on Sanctification.*[24] From the Reformed traditions, Anthony Hoekema describes holiness to function as "(1) separation from the sinful practices of this present world and (2) consecration to God's service."[25] And he continues, "Contrary to popular opinion, therefore, holiness means more than doing certain good things and not doing certain bad things; rather, it means being totally dedicated to God and separated from all that is sinful." This clarification regarding God's active role in nurturing holiness is not altogether different from the Wesleyan perspective, but God's providence is more strongly emphasized. While the Wesleyan traditions see holiness as a God-initiated movement followed by an individual response to grace, the Reformed traditions largely emphasize the former over the latter.[26] This nuance

---

[20]Randy L. Maddox, "John Wesley and Eastern Orthodoxy: Influences, Convergences, and Differences," *Asbury Theological Journal* 45, no. 2 (1990): 30.

[21]Ibid. A footnote (n4) from Maddox: "For three typical listings, see *Plain Account of Genuine Christianity*, §III.11, *John Wesley*, 195; 'An Address to the Clergy,' §§I.2 & II.1, *Works* (Jackson) 10:484, 492; and Sermon 112, 'On Laying the Foundation of the New Chape,l [sic]' §II.3, *Works* 3:586. Note also that Wesley included excerpts for Clement, Ignatius, Macarius and Polycarp in Volume 1 of his *Christian Library*."

[22]Ibid., 31.

[23]Christoforos Stavropoulos, "Partakers of Divine Nature," in *Eastern Orthodox Theology: A Contemporary Reader*, ed. D. B. Clendenin (Grand Rapids: Baker Publishing Group, 2003).

[24]Stanely N. Gundry, ed., *Five Views on Sanctification* (Grand Rapids: Zondervan, 1987).

[25]Anthony A. Hoekema, "The Reformed Perspective," in Gundry, *Five Views on Sanctification*.

[26]Yet Hoekema would concede that sanctification is still a both/and, where God's working meets humanity's effort. "Whose work is sanctification? In looking at the pattern of sanctification, we have observed that sanctification is both the world of God and the responsibility of His people.

proves significant when trying to understand where biological influences enter the mix. While it is clearly possible for individuals holding to Reformed theology to connect evolution and holiness, problems such as biological predispositions (that border on determinism) might be more difficult to overcome.

All in all, the main issues connecting holy living to our evolutionary influences are not altogether different from denomination to denomination. Any Christian tradition that takes seriously the call to holiness will have the same biological and environmental hurdles to overcome. Every tradition—even Reformed traditions that emphasize predestination—takes seriously the biblical commands to both struggle against the influences of this world, and the negative influences within ourselves.

7.3 PRACTICAL IMPLICATIONS AND FURTHER EXPLORATIONS

Besides certain theoretical nuances that are developed in this book, there are some practical implications as well. One such practical implication—having studied both biological altruism and Wesleyan holiness (which necessarily includes "outward" holiness) and concluding that there are numerous tangible suggestions of implementation—is that implementations center on the virtue of generosity and the church. As Stephen Pope suggests, generosity and grace "restores, purifies, and heals our natural ability to recognize in every person the goodness, worth, and dignity that they possess simply in virtue of being human."[27] Viewing the world in such a way leads to a myriad of practical implications that include service

---

Scripture plainly teaches that God is the author of sanctification. The work of sanctification, however, is ascribed to all three persons of the Trinity. . . . It is important for us to realize that sanctification is not something that we do by ourselves, with our own efforts and in our own strength. Sanctification is not a human activity but a divine gift. . . . Since we are God's covenant people, Paul is saying, we have a solemn responsibility. We must fight against sin, both of the body and of the mind. The Greek word *epitelountes*, translated 'perfecting,' has in it the root of the word for 'goal' (*telos*) and means 'progressively bringing to its goal.' What we usually think of as God's work is here vividly described as the believer's task, namely, to bring holiness to its goal." Ibid., 68-70.

[27] Pope, *Human Evolution and Christian Ethics*, 229. He goes on to say, "This reliance on grace might be taken to suggest that scientifically established knowledge of human nature is irrelevant to Christian ethics, but this is far from being the case because grace acts on and within human nature."

to the poor and charity to the needy. The crux of moral development and education, especially within the church, is the integrity of character and the connection to such virtues.[28]

In their book *Passing the Plate: Why American Christians Don't Give Away More Money*,[29] Christian Smith, Michael Emerson and Patricia Snell suggest that American churches in particular (although many of the claims could be extended to most churches in developed nations) have the wrong kind of community that ends up reducing altruism and generosity. In other words, the constraints that reside within these churches set up a community by which generosity and other-centered care is significantly discouraged.[30] Solutions to these problems are often multidimensional and include both lay and clergy cooperation.[31] According to the authors, some common reasons why most Christians do not give more generously are that they are not confronted with selfishness and its moral implications; Christians have low expectations from churches and communities; Christians do not often trust the church to handle issues of generosity; and there are no tangible consequences for stinginess.[32] All of these reasons, and no doubt more, result in "occasional generosity" rather than "lifestyle generosity."

With any book, there is no end to what one might explore. As with any major question that is asked, an answer ends up giving the questioner many more questions to be answered. So it is with the subject I attempted to address in this book. When looking at the significance of

---

[28]See Craig Steven Titus, *The Psychology of Character and Virtue* (Arlington, VA: Institute for the Psychological Sciences Press, 2009), 120-21. It should be noted that Wesleyan bands are classes that are not completely unique in their ability to create more altruistic people. There are other communities that nurture altruism. Altruism-creating environments need not be distinctly Christian or even, more broadly, religious, and could come about in political or social movements more generally. Yet, what makes Wesleyan bands and classes highly potent in nurturing altruistic behavior is the common religious motivation to become holy. This religious foundation makes Wesleyan groups more easily ready to engender altruism.

[29]Christian Smith, Michael O. Emerson and Patricia Snell, *Passing the Plate: Why American Christians Don't Give Away More Money* (Oxford: Oxford University Press, 2008).

[30]The thesis of Robert Wuthnow's book, *God and Mammon in America*, also supports the claim that organized religion in America actually encourages rampant materialism. See Wuthnow, *God and Mammon in America* (Toronto: Free Press, 1994).

[31]Smith, Emerson and Snell, *Passing the Plate*, 147. What is interesting to note is that the higher the household income the less one actually gives: $10,000/y = 2.3 percent and $70,000/y = 1.2%. See ibid., 44.

[32]Ibid., 97.

212 EVOLUTION AND HOLINESS

sociobiology for Wesleyan ethics, there were many side projects that, while wholly worthwhile for future research, were beyond the scope of my book.

If time and space had allowed, one interesting subject matter worth exploring would have been what the intersection of *theosis*, genetics and the "life to come" might look like. For instance, lions do not have the genetic proclivity to lie down with lambs. In the long run, orthodox Christian teaching holds that God will restore all of creation, either returning it to a state before the fall (in the Augustinian traditions) or helping it reach perfection (in the Eastern Orthodox traditions). The question remains, what happens, for example, to organisms genetically composed to be bent toward violence? How do those organisms' genes function in this new earth?

Another avenue of further exploration would be the idea that being good or altruistic is a constant battle. This idea is often championed by those who cite the "inner conflict" sections of Romans 7:14-25.[33] Some questions concern whether goodness is a constant struggle; what it might mean to have overcome our genetics, since people are not less than fully bodily; and, in some sense, how sin is related to the will rather than to the involuntary or merely biological (i.e., a purely physical involuntary reflex is a movement, not an action). Yet, as Ben Witherington III observes, this passage has been completely misconstrued by many Christians.[34] He says that the *I* in verses 14-25 must be seen as fictive because Paul's argument is about the "first Adam" and his fallen nature, noting further that "Paul is not talking about the Christian who tries hard but

---

[33]Romans 7:14-25: "For we know that the law is spiritual; but I am of the flesh, sold into slavery under sin. I do not understand my own actions. For I do not do what I want, but I do the very thing I hate. Now if I do what I do not want, I agree that the law is good. But in fact it is no longer I that do it, but sin that dwells within me. For I know that nothing good dwells within me, that is, in my flesh. I can will what is right, but I cannot do it. For I do not do the good I want, but the evil I do not want is what I do. Now if I do what I do not want, it is no longer I that do it, but sin that dwells within me. So I find it to be a law that when I want to do what is good, evil lies close at hand. For I delight in the law of God in my inmost self, but I see in my members another law at war with the law of my mind, making me captive to the law of sin that dwells in my members. Wretched man that I am! Who will rescue me from this body of death? Thanks be to God through Jesus Christ our Lord! So then, with my mind I am a slave to the law of God, but with my flesh I am a slave to the law of sin."

[34]See Ben Witherington and Darlene Hyatt, *Paul's Letter to the Romans: A Socio-Rhetorical Commentary* (Grand Rapids: Eerdmans, 2004), 193.

whose deeds do not match up with his intentions. The point here is not the falling short or imperfection even of Christian good deeds, but the exceeding sinfulness of sinful deeds."[35] In this passage, Paul is not claiming that he *himself* constantly does what he knows to be wrong (with a kind of inevitability); rather, it is Paul *as the first Adam* in a fallen state who struggles. What is more, in Christ—the second Adam—the Christian can be more than a conqueror, as we see in Romans 8. This message, taken in view of the context of the entire book of Romans, would have been evident to early listeners and readers of the text. Consequently, under this interpretation of Romans 7, an individual would not be "bound" by her or his genetic material, making the connection of Romans 7 to genetics and ethics a landscape for further research.

In chapter five, I developed several theological underpinnings supporting why Wesley developed his highly structured bands and classes. The foundation of those underpinnings was the concept of *theosis*, which Wesley borrowed from Eastern Orthodoxy. One limitation for Wesley was that he accepted the framework of Augustine on original sin *and* the Eastern Orthodox framework on the *imago Dei* and perfection. This relationship created a tension for Wesley. In light of this tension, a deeper exploration of his concept of creation and the fall would be merited—though such is beyond the range of this project.[36] To the extent that he believed humanity was created in an original state of perfection, subsequently fell away from that state and finally hoped to regain perfection, Wesley kept in step with Western Christianity.[37] To the extent that Wesley leaned toward the East, viewing Adam as not perfect but rather underdeveloped, he was more compatible with sociobiological explanations of altruism. To the extent that Wesley leaned toward an Augustinian model of creation and the fall, seeing Adam in a perfect state that he fell away from, Wesley was less compatible with sociobiological explanations of altruism. While I certainly think Wesley's position, leaning

---

[35]Ibid., 200.

[36]Wesley thought humanity fell away from perfection (with Adam) and then attempts to go back to a state of perfection. See the sermon titled "The Image of God," in Wesley, *Sermons 115–151*, 290–303. See also Randy L. Maddox, *Responsible Grace: John Wesley's Practical Theology* (Nashville: Kingswood Books, 1994), 67.

[37]Maddox, *Responsible Grace*, 67.

more toward Augustinian but tempered with Eastern Orthodox thinking, is not antithetical to sociobiology, additional research could be done into how Wesleyanism could have benefited if its founder moved more toward the East.[38]

Another aspect that could prove helpful in this discussion comes from Ian Barbour. In the book *When Science Meets Religion*, Barbour provides a helpful paradigm for the intersection of science and religion, illustrating four ways these disciplines can interact: conflict, independence, dialogue and integration.[39] Again, for Wesley, the field of sociobiology was not around when he was alive, making reading Wesley with sociobiology always an after-the-fact exercise. Nevertheless, from what we know of Wesleyan theology, we could likely place it in Barbour's "dialogue" phase with sociobiology, though not quite in the "integration" phase.[40] In this way, we can see similar parallels between the fields of sociobiology and Wesleyan theology.[41]

A prevalent presupposition that is present in both evolutionary biology and Wesleyan theology would be the acknowledgement that human nature has a selfish side. For Wesley, there was no good in the "natural human" apart from the prevenient grace of God. It just so happens that Wesley also believed, as noted in chapter five, that *all* people had the prevenient grace of God working in their lives. Similarly, sociobiologists and evolutionary biologists have long argued that human nature was selfish—ever promoting its individual agenda. However, as

---

[38]John Hick could prove helpful here. See especially the chapter titled "A Philosophy of Religious Pluralism" in Hick, *The New Frontier of Religion and Science: Religious Experience, Neuroscience and the Transcendent* (New York: Palgrave Macmillan, 2010), 162-71.

[39]See chapter one of Ian G. Barbour, *When Science Meets Religion* (San Francisco: HarperSanFrancisco, 2000).

[40]Markus Mühling says, "Natural science and theology can only fruitfully lead a dialogue if philosophy and especially natural philosophy are taken into account. Therefore, there can be no dialogue without a 'trialogue.'" *Resonances: Neurobiology, Evolution and Theology; Evolutionary Niche Construction, the Ecological Brain and Relational-Narrative Theology* (Göttingen: Vandenhoeck & Ruprecht, 2014), 225.

[41]One can also see how connecting the fields of sociobiology and holiness might develop a bridge between seemingly disparate groups. There is already some hope—if even small—in this connection in the comments from David Sloan Wilson: "I admire a good religion for its ability to motivate altruism at the level of action, in the same way that I admire the wisdom of a multicellular organism or a beehive. I hope that my intellectual brethren and I can do as well." Wilson, *Does Altruism Exist? Culture, Genes, and the Welfare of Others*, Foundational Questions in Science (New Haven, CT: Yale University Press and Templeton Press, 2015), 91.

of late, there has been much discussion concerning genetic cooperation among various organisms.[42] It seems that these two fields have drawn similar conclusions. Yet I do not think there is enough evidence to show that Wesleyan theology could be in Barbour's "integration" phase with sociobiology. For his theology to be considered to be in such a phase, a systematic synthesis would need to occur between the two fields—and Wesley simply does not lean far enough away from Augustine for this to happen. However, if Wesley had gone further, leaning even more on the Eastern Orthodox tradition rather than Augustinian traditions, where he would have adopted an Eastern framework of not just *theosis* but also creation and fall, his theology would more seamlessly integrate with sociobiology. I attribute this lack of integration to the fact that Wesley lived well before the fields of either sociobiology or evolutionary biology existed. Although it is beyond the purview of this research, studying how Wesleyan theology is compatible and can be *fully* "integrated" with sociobiology would be a worthwhile project on which to embark.[43]

7.4 CONCLUSION

Human beings are wedged between their genetic proclivities and their environmental constraints. While influences come from all sides and in different ways, humans have the uncanny ability to shift how those influences sway their behavior. Within Wesleyan ethics, this ability is planted in free will and grows to fruition in the grace of God. A wonderful example of this relationship can be seen in John Wesley's own life. As Oden writes,

> Wesley did not have a passive, idle, lethargic, quietistic notion of saving grace. Its reception requires energetic work, earnest prayer, spirited study of Scripture, and active good works. It is not as if God zaps us with grace apart

---

[42]This can be seen in multilevel selection theory (or group selection theory).

[43]Also beyond the scope of this research is the idea of natural law compatibility with sociobiology, which seems to provide a good deal of evidence for drives for self-preservation that fit nicely with Thomistic natural law morality. It might even be said that sociobiology is compatible with some version of natural law morality, though there are some areas of natural law that transcend pure sociobiological explanations. For more on this, see Craig A. Boyd, "Thomistic Natural Law and the Limits of Evolutionary Psychology," in *Evolution and Ethics: Human Morality in Biological and Religious Perspective*, ed. Philip Clayton and Jeffrey Schloss (Grand Rapids: Eerdmans, 2004), 235.

from our responsive cooperation. Every subsequent act of cooperating with grace is premised on God's preceding grace, which elicits and requires free human responsiveness.[44]

The individual cannot go about expecting to live a life of little action, letting the grace of God force him into a life of holiness with outward altruistic action. Nor can the individual expect to obtain this perfected life wholly according to her or his own free will. In some enigmatic way, much like many teachings of the Christian church, humans shift on the selfish/selfless spectrum as a result of the mysterious interplay between free will, constraints on individuals—both biological and environmental—and free grace from God.

For these reasons, the Christian individual wishing to develop a life of holiness with altruistic action as fruit must reside within a community where a system of constraints encourages such behavior—just as Wesley displayed with his bands and classes. These concrete communities provide conceptions of the human good to be pursued through the particular virtues that they uphold as central to their notion of what is good.[45] These structures are not necessarily burdensome, but rather provide a conception of the human good for those who participate in the defining practices of its way of life.[46] Wesley's own example is very helpful in illustrating the dynamic action combined with unmerited grace that is required in order to move into holiness. As Mildred Bangs Wynkoop says,

> His lifelong search for perfection constitutes the secret of Wesley's temper.
> This is not to be interpreted as a fruitless, failing quest of an ever receding
> "will-o-the-wisp." Nothing could be farther from the truth. But he was a "file
> leader" in religion because he never rested in the achievement of the moment.
> The very nature of the Christian life is progress. Perfection is not a static

---

[44]Oden, *John Wesley's Teachings: Christ and Salvation*, 142.

[45]Alasdair C. MacIntyre, *After Virtue: A Study in Moral Theory*, 3rd ed. (Notre Dame, IN: University of Notre Dame Press, 2007), 58.

[46]Ibid. In a similar way, regarding the connection between evolutionary theory and Christian community, Stephen Pope says that "Christian narrative and participation within the Christian community supply a context for interpreting human evolution, yet the latter gave rise to capacities and inclinations developed within the Christian life. There is a kind of circulatory, but not a vicious one, in the relation between faith and nature in this regard." Pope, *Human Evolution and Christian Ethics*, 296.

"having" but a dynamic "going." Love is not "perfect" in the sense of having reached its zenith, but in its quality as a dynamic relationship subject to infinite increase.[47]

Consequently, it is the church's task to do the same—ever leaning on the grace of God while working to develop a lifestyle of holiness that bears altruistic fruit.

---

[47]Mildred Bangs Wynkoop, *A Theology of Love: The Dynamic of Wesleyanism* (Kansas City, MO: Beacon Hill Press, 1972), 66.

EXCERPT FROM "PRINCIPLES OF A METHODIST" ON THE TOPIC OF CHRISTIAN PERFECTION

1. Perhaps the general prejudice against Christian perfection may chiefly arise from a misapprehension of the nature of it. We willingly allow, and continually declare, there is no such perfection in this life, as implies either a dispensation from doing good, and attending all the ordinances of God, or a freedom from ignorance, mistake, temptation, and a thousand infirmities necessarily connected with flesh and blood.

2. First. We not only allow, but earnestly contend, that there is no perfection in this life, which implies any dispensation from attending all the ordinances of God, or from doing good unto all men while we have time, though "especially unto the household of faith." We believe, that not only the babes in Christ, who have newly found redemption in his blood, but those also who are "grown up into perfect men," are indispensably obliged, as often as they have opportunity, "to eat bread and drink wine in remembrance of Him," and to "search the Scriptures;" by fasting, as well as temperance, to "keep their bodies under, and bring them into subjection;" and, above all, to pour out their souls in prayer, both secretly, and in the great congregation.

"Principles of a Methodist" published in John Wesley, *A Plain Account of Christian Perfection* (New York: Land & Scott, 1850), 40-44. Randy Maddox notes that when reading this Wesleyan scholars consistently identify inward holiness with the Christian tempers (or habits). See Randy L. Maddox, *Responsible Grace: John Wesley's Practical Theology* (Nashville: Kingswood Books, 1994), 289n34.

3. We Secondly believe, that there is no such perfection in this life, as implies an entire deliverance, either from ignorance, or mistake, in things not essential to salvation, or from manifold temptations, or from numberless infirmities, wherewith the corruptible body more or less presses down the soul. We cannot find any ground in Scripture to suppose, that any inhabitant of a house of clay is wholly exempt either from bodily infirmities, or from ignorance of many things; or to imagine any is incapable of mistake, or falling into divers temptations.

4. But whom then do you mean by "one that is perfect?" We mean one in whom is "the mind which was in Christ," and who so "walketh as Christ also walked;" a man "that hath clean hands and a pure heart," or that is "cleansed from all filthiness of flesh and spirit;" one in whom is "no occasion of stumbling," and who, accordingly, "does not commit sin." To declare this a little more particularly: We understand by that scriptural expression, "a perfect man," one in whom God hath fulfilled his faithful word, "From all your filthiness and from all your idols I will cleanse you: I will also save you from all your uncleannesses." We understand hereby, one whom God hath "sanctified throughout in body, soul, and spirit;" one who "walketh in the light as He is in the light, in whom is no darkness at all; the blood of Jesus Christ his Son having cleansed him from all sin."

5. This man can now testify to all mankind, "I am crucified with Christ: Nevertheless I live; yet not I, but Christ liveth in me." He is "holy as God who called" him "is holy," both in heart and "in all manner of conversation." He "loveth the Lord his God with all his heart," and serveth him "with all his strength." He "loveth his neighbour," every man, "as himself;" yea, "as Christ loveth us;" them, in particular, that "despitefully use him and persecute him, because they know not the Son, neither the Father." Indeed his soul is all love, filled with "bowels of mercies, kindness, meekness, gentleness, longsuffering." And his life agreeth thereto, full of "the work of faith, the patience of hope, the labour of love." "And whatsoever" he "doeth either in word or deed," he "doeth it all in the name," in the love and power, "of the Lord

Jesus." In a word, he doeth "the will of God on earth, as it is done in heaven."

6. This it is to be a perfect man, to be "sanctified throughout;" even "to have a heart so all-flaming with the love of God," (to use Archbishop Usher's words,) "as continually to offer up every thought, word, and work, as a spiritual sacrifice, acceptable to God through Christ." In every thought of our hearts, in every word of our tongues, in every work of our hands, to "show forth his praise, who hath called us out of darkness into his marvellous light." O that both we, and all who seek the Lord Jesus in sincerity, may thus "be made perfect in one!"

EXCERPT FROM "A PLAIN ACCOUNT OF CHRISTIAN PERFECTION"

In the year 1764, upon a review of the whole subject, I wrote down the sum of what I had observed in the following short propositions:

1. There is such a thing as perfection; for it is again and again mentioned in Scripture.

2. It is not so early as justification; for justified persons are to "go on unto perfection" (Heb. vi. i).

3. It is not so late as death; for St. Paul speaks of living men that were perfect (Phil. iii. 15).

4. It is not absolute. Absolute perfection belongs not to man, nor to angels, but to God alone.

5. It does not make a man infallible: None is infallible, while he remains in the body.

6. Is it sinless? It is not worth while to contend for a term. It is "salvation from sin."

7. It is "perfect love" (1 John v. 18). This is the essence of it: its properties, or inseparable fruits, are, rejoicing evermore, praying without ceasing, and in everything giving thanks (1 Thess. v. 16, etc.).

8. It is improvable. It is so far from lying in an indivisible point, from

John Wesley, *A Plain Account of Christian Perfection* (New York: Land & Scott, 1850), 166-69.

being incapable of increase, that one perfected in love may grow in grace far swifter than he did before.

9. It is amissible, capable of being lost; of which we have numerous instances. But we were not thoroughly convinced of this till five or six years ago.

10. It is constantly both preceded and followed by a gradual work.

11. But is it in itself instantaneous or not? In examining this, let us go on step by step.

An instantaneous change has been wrought in some believers: None can deny this.

Since that change, they enjoy perfect love; they feel this and this alone; they "rejoice evermore, pray without ceasing and in everything give thanks." Now, this is all that I mean by perfection; therefore, these are witnesses of the perfection which I preach.

"But in some this change was not instantaneous?" They did not perceive the instant when it was wrought. It is often difficult to perceive the instant when a man dies yet there is an instant in which life ceases. And if ever sin ceases, there must be a last moment of its existence, and a first moment of our deliverance from it.

"But if they have this love now, they will lose it?" They may; but they need not. And whether they do or no, they have it now; they now experience what we teach. They now are all love; they now rejoice, pray, and praise without ceasing.

"However, sin is only suspended in them; it is not destroyed?" Call it which you please. They are all love to-day; and they take no thought for the morrow.

"But this doctrine has been much abused." So has that of justification by faith. But that is no reason for giving up either this or any other scriptural doctrine. "When you wash your child," as one speaks, "throw away the water but do not throw away the child."

"But those who think they are saved from sin say they have no need of the merits of Christ." They say just the contrary. Their language is:

Every moment, Lord I want the merit of thy death!

They never before had so deep, so unspeakable, a conviction of the need of Christ in all his offices as they have now.

Therefore, all our Preachers should make a point of preaching perfection to believers constantly, strongly, and explicitly; and all believers should mind this one thing, and continually agonize for it.

RULES OF THE BAND SOCIETIES
DRAWN UP DEC. 25, 1738

The design of our meeting is, to obey that command of God, "Confess your faults one to another, and pray one for another, that ye may be healed."

To this end, we intend,

1. To meet once a week, at the least.

2. To come punctually at the hour appointed, without some extraordinary reason.

3. To begin (those of us who are present) exactly at the hour, with singing or prayer.

4. To speak, each of us in order, freely and plainly the true state of our souls, with the faults we have committed in thought, word, or deed, and the temptations we have felt since our last meeting.

5. To end every meeting with prayer, suited to the state of each person present.

6. To desire some person among us to speak his own state first, and then to ask the rest in order as many and as searching questions as may be concerning their state, sins, and temptations.

Some of the questions proposed to every one before he is admitted amongst us may be to this effect:

1. Have you the forgiveness of your sins?

From John Wesley, *The Works of the Reverend John Wesley, A. M.*, 1st American ed. (New York: J. Emory and B. Waugh, 1831), 5:192-94.

2. Have you peace with God, through our Lord Jesus Christ?

3. Have you the witness of God's Spirit with your spirit, that you are a child of God?

4. Is the love of God shed abroad in your heart?

5. Has no sin, inward or outward, dominion over you?

6. Do you desire to be told of your faults?

7. Do you desire to be told of all your faults, and that plain and home?

8. Do you desire that every one of us should tell you from time to time whatsoever is in his heart concerning you?

9. Consider! Do you desire we should tell you whatsoever we think, whatsoever we fear, whatsoever we hear, concerning you?

10. Do you desire that in doing this we should come as close as possible, that we should cut to the quick, and search your heart to the bottom?

11. Is it your desire and design to be on this and all other occasions entirely open, so as to speak everything that is in your heart, without exception, without disguise, and without reserve?

Any of the preceding questions may be asked as often as occasion offers; the five following at every meeting:

1. What known sins have you committed since our last meeting?

2. What temptations have you met with?

3. How were you delivered?

4. What have you thought, said, or done, of which you doubt whether it be sin or not?

5. Have you nothing you desire to keep secret?

DIRECTIONS GIVEN TO THE BAND SOCIETIES[1]
DEC. 25, 1744

You are supposed to have the faith that "overcometh the world." To you, therefore, it is not grievous,

---

[1]Ibid., 79.

I. Carefully to abstain from doing evil; in particular,

    1. Neither to buy nor sell anything at all on the Lord's day.

    2. To taste no spirituous liquor, no dram of any kind, unless prescribed by a physician.

    3. To be at a word both in buying and selling.

    4. To pawn nothing, no, not to save life.

    5. Not to mention the fault of any behind his back, and to stop those short that do.

    6. To wear no needless ornaments, such as rings, earrings, necklaces, lace, ruffles.

    7. To use no needless self-indulgence, such as taking snuff or tobacco, unless prescribed by a physician.

II. Zealously to maintain good works; in particular,

    1. To give alms of such things as you possess, and that to the uttermost of your power.

    2. To reprove all that sin in your sight, and that in love and meekness of wisdom.

    3. To be patterns of diligence and frugality, of self-denial, and taking up the cross daily.

III. Constantly to attend on all the ordinances of God; in particular,

    1. To be at church and at the Lord's table every week, and at every public meeting of the bands.

    2. To attend the ministry of the word every morning unless distance, business, or sickness prevent.

    3. To use private prayer every day, and family prayer if you are the head of a family.

    4. To read the Scriptures, and meditate thereon, at every vacant hour. And,

    5. To observe as days of fasting or abstinence all Fridays in the year.

BIBLIOGRAPHY

À Kempis, Thomas. *The Imitation of Christ*. Edited by Aloysius Croft and Harold Bolton. Dover Thrift Editions. Mineola, NY: Dover Publications, 2003.

Abraham, William J. *Aldersgate and Athens: John Wesley and the Foundations of Christian Belief*. Waco, TX: Baylor University Press, 2010.

Alexander, Richard D. *The Biology of Moral Systems*. Foundations of Human Behavior. Hawthorne, NY: A. de Gruyter, 1987.

———. *Darwinism and Human Affairs*. The Jessie and John Danz Lectures. Seattle: University of Washington Press, 1979.

———. "The Search for a General Theory of Behavior." *Behavioral Science* 20, no. 2 (1975): 77-100.

Appleman, Philip. *Darwin*. 2nd ed. A Norton Critical Edition. New York: Norton, 1979.

Aquinas, Thomas. *Summa Theologica*. Translated by Fathers of the English Dominican Province. New York: Benziger, 1947-48. Reprint ed., Allen, TX: Christian Classics, 1981.

Aristotle. *Nicomachean Ethics*. Translated by Martin Ostwald. The Library of Liberal Arts 75. Indianapolis: Bobbs-Merrill, 1962.

Arnhart, Larry. *Darwinian Natural Right: The Biological Ethics of Human Nature*. SUNY Series in Philosophy and Biology. Albany: State University of New York Press, 1998.

———. "Thomistic Natural Law as Darwinian Natural Right." *Social Philosophy and Policy* 18, no. 1 (2001): 1-33.

Axelrod, Robert M. *The Evolution of Cooperation*. New York: Basic Books, 1984.

Ayala, Francisco. "The Difference of Being Human." In *Biology, Ethics, and the Origins of Life*, edited by Holmes Rolston. The Jones and Bartlett Series in Philosophy. Boston: Jones and Bartlett, 1995.

Bangs, Carl. *Our Roots of Belief: Bible Faith and Faithful Theology*. Kansas City, MO: Beacon Hill Press, 1981.

Barbour, Ian G. *When Science Meets Religion*. San Francisco: HarperSanFrancisco, 2000.

Barlow, George W., and James Silverberg. *Sociobiology, Beyond Nature/Nurture? Reports, Definitions, and Debate*. Boulder, CO: Westview Press for the American Association for the Advancement of Science, 1980.

Bate, John, ed. *The Local Preacher's Treasury*. London: Lile and Fawcett, 1884.

Batson, C. Daniel. *The Altruism Question: Toward a Social Psychological Answer*. Hillsdale, NJ: L. Erlbaum, Associates, 1991.

Bebb, E. Douglas. *Wesley: A Man with a Concern*. London: Epworth Press, 1950.

Beeman, R. W., K. S. Friesen and R. E. Denell. "Maternal-Effect Selfish Genes in Flour Beetles." *Science* 256, no. 5053 (1992): 89-92.

Berkhof, Louis. *Systematic Theology*. Grand Rapids: Eerdmans, 1996.

Berry, Wendell. *Life Is a Miracle: An Essay Against Modern Superstition*. Washington, DC: Counterpoint, 2000.

Blackmore, Susan J. *The Meme Machine*. Oxford: Oxford University Press, 1999.

Boehm, Christopher. *Hierarchy in the Forest: The Evolution of Egalitarian Behavior*. Cambridge, MA: Harvard University Press, 1999.

Bowker, John. *Is God a Virus? Genes, Culture, and Religion*. The Gresham Lectures. London: SPCK, 1995.

———. "Origins, Functions, and Management of Aggression in Biocultural Evolution." *Zygon* 18 (1983).

Boyd, Craig A. "Thomistic Natural Law and the Limits of Evolutionary Psychology." In *Evolution and Ethics: Human Morality in Biological and Religious Perspective*, edited by Philip Clayton and Jeffrey Schloss, 221-37. Grand Rapids: Eerdmans, 2004.

Bråten, Stein. *On Being Moved: From Mirror Neurons to Empathy*. Advances in Consciousness Research. Amsterdam: John Benjamins, 2007.

Bready, J. Wesley. *England: Before and After Wesley*. London: Hodder and Stoughton, 1938.

Buckley, Sarah J. *Gentle Birth, Gentle Mothering: A Doctor's Guide to Natural Childbirth and Gentle Early Parenting Choices*. Berkeley, CA: Celestial Arts, 2009.

Buscemi, L., and C. Turchi. "An Overview of the Genetic Susceptibility to Alcoholism." *Medicine, Science, and the Law* 51 (2011): 2-6.

Campolo, Anthony. *A Reasonable Faith: Responding to Secularism*. Waco, TX: Word Books, 1983.

Charlesworth, Brian, and Daniel L. Hartl. "Population Dynamics of the Segregation Distorter Polymorphism of Drosophila Melanogaster." *Genetics* 89, no. 1 (May 1978): 171-92.

Chilcote, Paul Wesley. *Recapturing the Wesleys' Vision: An Introduction to the Faith of John and Charles Wesley*. Downers Grove, IL: InterVarsity Press, 2004.

Christensen, Michael J. "Theosis and Sanctification: John Wesley's Reformulation of a Patristic Doctrine." *Wesley Theological Journal* 31, no. 2 (1996): 71-94.

Churchland, Patricia S. *Braintrust: What Neuroscience Tells Us About Morality.* Princeton, NJ: Princeton University Press, 2011.

Clark, Stephen R. L. *Biology and Christian Ethics.* Cambridge: Cambridge University Press, 2000.

Clayton, Philip. "Biology and Purpose: Altruism, Morality, and Human Nature in Evolutionary Perspective." In *Evolution and Ethics: Human Morality in Biological and Religious Perspective*, edited by Philip Clayton and Jeffrey Schloss, 318-36. Grand Rapids: Eerdmans, 2004.

————. "Evolution, Altruism, and God: Why the Levels of Emergent Complexity Matter." In *Evolution, Games, and God: The Principle of Cooperation*, edited by Martin A. Nowak and Sarah Coakley. Cambridge, MA: Harvard University Press, 2013.

Clayton, Philip, and Jeffrey Schloss, eds. *Evolution and Ethics: Human Morality in Biological and Religious Perspective.* Grand Rapids: Eerdmans, 2004.

Coakley, Sarah. *Sacrifice Regained: Evolution, Cooperation, and God.* Gifford Lectures. University of Aberdeen, 2012. http://www.abdn.ac.uk/gifford/about/.

Collins, Kenneth J. *John Wesley: A Theological Journey.* Nashville: Abingdon Press, 2003.

————. *The Scripture Way of Salvation: The Heart of John Wesley's Theology.* Nashville: Abingdon Press, 1997.

————. *The Theology of John Wesley: Holy Love and the Shape of Grace.* Nashville: Abingdon Press, 2007.

Corradini, A., S. Galvan and E. J. Lowe. *Analytic Philosophy Without Naturalism.* London: Routledge, 2006.

Crick, Francis. *The Astonishing Hypothesis: The Scientific Search for the Soul.* New York: Maxwell Macmillan International, 1994.

Cronk, Lee, and Beth L. Leech. *Meeting at Grand Central: Understanding the Social and Evolutionary Roots of Cooperation.* Princeton, NJ: Princeton University Press, 2012.

Cudworth, Ralph. *The Life of Christ the Pith and Kernel of All Religion: A Sermon Preached Before the Honourable House of Commons, at Westminster, March 31, 1647.* Westminster: BiblioBazaar, 2010.

Cunningham, Conor. *Darwin's Pious Idea: Why the Ultra-Darwinists and Creationists Both Get It Wrong.* Grand Rapids: Eerdmans, 2010.

Darwin, Charles. *The Darwin Reader*. Edited by Mark Ridley. New York: Norton, 1987.

———. *The Descent of Man, and Selection in Relation to Sex*. 2 vols. New York: D. Appleton and Company, 1871.

———. *The Descent of Man and Selection in Relation to Sex*. 2nd ed. New York: P. F. Collier, 1902.

———. *The Descent of Man and Selection in Relation to Sex*. New York: Penguin, 2004.

———. *The Origin of Species*. The Harvard Classics. New York: Collier, 1961.

Dawkins, R., and J. R. Krebs. "Arms Races Between and Within Species." *Proceedings of the Royal Society of London: Series B, Biological Sciences* 205, no. 1161 (1979): 489-511.

Dawkins, Richard. *The Blind Watchmaker*. New York: Norton, 1986.

———. *The Extended Phenotype: The Gene as the Unit of Selection*. Oxford: Freeman, 1982.

———. *River out of Eden: A Darwinian View of Life*. Science Masters Series. New York: Basic Books, 1995.

———. *The Selfish Gene*. New York: Oxford University Press, 1976.

———. *The Selfish Gene*. New ed. Oxford: Oxford University Press, 1989.

———. "Selfish Genes and Selfish Memes." In *The Mind's I: Fantasies and Reflections on Self and Soul*, edited by Douglas R. Hofstadter and Daniel Clement Dennett. New York: Basic Books, 1981.

De Waal, Frans. *The Age of Empathy: Nature's Lessons for a Kinder Society*. New York: Harmony Books, 2009.

———. *Good Natured: The Origins of Right and Wrong in Humans and Other Animals*. Cambridge, MA: Harvard University Press, 1996.

De Waal, Frans, Stephen Macedo, Josiah Ober and Robert Wright, eds. *Primates and Philosophers: How Morality Evolved*. The University Center for Human Values Series. Princeton, NJ: Princeton University Press, 2006.

Deane-Drummond, Celia. *Christ and Evolution: Wonder and Wisdom*. Minneapolis: Fortress Press, 2009.

———. *Creation Through Wisdom: Theology and the New Biology*. Edinburgh: T&T Clark, 2000.

———. *The Ethics of Nature*. New Dimensions to Religious Ethics. Malden, MA: Blackwell, 2004.

Deane-Drummond, Celia, Bronislaw Szerszynski and Robin Grove-White. *Re-Ordering Nature: Theology, Society, and the New Genetics*. London: T&T Clark, 2003.

Dennett, Daniel C. "Animal Consciousness: What Matters and Why." *Social Research* 62, no. 3 (1995): 691-710.

———. *Brainstorms: Philosophical Essays on Mind and Psychology*. Montgomery, VT: Bradford Books, 1978.

———. *Consciousness Explained*. Boston: Little, Brown and Co., 1991.

———. *Darwin's Dangerous Idea: Evolution and the Meanings of Life*. New York: Simon & Schuster, 1995.

———. *Elbow Room: The Varieties of Free Will Worth Wanting*. Cambridge, MA: MIT Press, 1984.

———. *Freedom Evolves*. New York: Viking, 2003.

———. *Kinds of Minds: Toward an Understanding of Consciousness*. Science Masters. New York: Basic Books, 1996.

Descartes, René. *Discourse on Method: Meditations on First Philosophy*. Translated by Donald A. Cress. 3rd ed. Indianapolis: Hackett Pub. Co., 1993.

Diamond, Jared M. *The Third Chimpanzee: The Evolution and Future of the Human Animal*. New York: HarperCollins, 1992.

Dimond, Sydney George. *The Psychology of the Methodist Revival: An Empirical and Descriptive Study*. London: Oxford University Press, 1926.

Dobzhansky, Theodosius. "Nothing in Biology Makes Sense Except in the Light of Evolution." *American Biology Teacher* 35 (March 1973): 125-29.

Dover, Gabriel. "Anti-Dawkins." In *Alas, Poor Darwin: Arguments Against Evolutionary Psychology*, edited by Hilary Rose and Steven P. R. Rose. New York: Harmony Books, 2000.

Drees, Willem B. *Religion, Science, and Naturalism*. Cambridge: Cambridge University Press, 1996.

Dunnington, Kent. *Addiction and Virtue: Beyond the Models of Disease and Choice*. Strategic Initiatives in Evangelical Theology. Downers Grove, IL: IVP Academic, 2011.

Eagleman, David. "Incognito: What's Hiding in the Unconscious Mind." Interview with Terry Gross. *Fresh Air*. May 31, 2011. National Public Radio.

Erikson, Erik H. *Childhood and Society*. 2nd ed. New York: Norton, 1963.

Fischer, Eric A. "The Relationship Between Mating System and Simultaneous Hermaphroditism in the Coral Reef Fish, Hypoplectrus Nigricans (Serranidae)." *Animal Behaviour* 28, no. 2 (1980): 620-33.

Flanagan, Owen J. *The Science of the Mind*. 2nd ed. Cambridge, MA: MIT Press, 1991.

Fodor, Jerry A., and Massimo Piattelli-Palmarini. *What Darwin Got Wrong*. New York: Farrar, Straus and Giroux, 2010.

Frank, Robert H. *Passions Within Reason: The Strategic Role of the Emotions.* New York: Norton, 1988.

French, Peter A. *Responsibility Matters.* Lawrence: University Press of Kansas, 1992.

Ghiselin, Michael T. *The Economy of Nature and the Evolution of Sex.* Berkeley: University of California Press, 1974.

Goodwin, Donald W. *Alcoholism: The Facts.* 3rd ed. Oxford: Oxford University Press, 2000.

Gould, Stephen Jay. "More Things in Heaven and Earth." In *Alas, Poor Darwin: Arguments Against Evolutionary Psychology,* edited by Hilary Rose and Steven P. R. Rose. New York: Harmony Books, 2000.

———. *Punctuated Equilibrium.* Cambridge, MA: Belknap Press of Harvard University Press, 2007.

Gould, Stephen, and Niles Eldredge. "Punctuated Equilibria: The Tempo and Mode of Evolution Reconsidered." *Paleobiology* 3, no. 2 (1977): 115-51.

Green, Richard. *The Works of John and Charles Wesley: A Bibliography Containing an Exact Account of All the Publications Issued by the Brothers Wesley, Arranged in Chronological Order, with a List of the Early Editions, and Descriptive and Illustrative Notes.* 2nd ed. London: Methodist Publishing House, 1906.

Gundry, Stanley N., ed. *Five Views on Sanctification.* Grand Rapids: Zondervan, 1987.

Gunton, Colin E. "The Doctrine of Creation." In *The Cambridge Companion to Christian Doctrine,* edited by Colin E. Gunton, 141-57. Cambridge: Cambridge University Press, 1997.

Hamilton, W. D. "Altruism and Related Phenomena, Mainly in Social Insects." *Annual Review of Ecology and Systematics* 3, no. 1 (1972): 193-232.

———. "The Evolution of Altruistic Behavior." *The American Naturalist* 97, no. 896 (1963): 354.

———. "The Genetical Evolution of Social Behaviour 1." *Journal of Theoretical Biology* 7, no. 1 (1964): 1-16.

———. "The Genetical Evolution of Social Behaviour 2." *Journal of Theoretical Biology* 7, no. 1 (1964): 17-52.

Hamilton, W. D., and Mark Ridley. *Narrow Roads of Gene Land: Last Words.* Oxford: Oxford University Press, 2005.

Hauerwas, Stanley. *A Community of Character: Toward a Constructive Christian Social Ethic.* Notre Dame, IN: University of Notre Dame Press, 1981.

Henderson, D. Michael. *John Wesley's Class Meeting: A Model for Making Disciples.* Nappanee, IN: Evangel Pub. House, 1997.

Hick, John. *The New Frontier of Religion and Science: Religious Experience, Neuroscience and the Transcendent*. New York: Palgrave Macmillan, 2010.

Hobbes, Thomas. *Leviathan: Or, the Matter, Forme and Power of a Commonwealth, Ecclesiasticall and Civil*. Collier Classics in the History of Thought. New York: Collier Books, 1962.

Hoekema, Anthony A. "The Reformed Perspective." In *Five Views on Sanctification*, edited by Stanley N. Gundry, 59-102. Grand Rapids: Zondervan, 1987.

Hölldobler, Bert, and Edward O. Wilson. *The Superorganism: The Beauty, Elegance, and Strangeness of Insect Societies*. New York: W. W. Norton, 2009.

Huchingson, James Edward. *Religion and the Natural Sciences: The Range of Engagement*. Fort Worth, TX: Harcourt Brace Jovanovich College Publishers, 1993.

Hudson, W. D. *The Is-Ought Question: A Collection of Papers on the Central Problems in Moral Philosophy*. Controversies in Philosophy. London: Macmillan, 1969.

Huxley, Julian. *Evolution: The Modern Synthesis*. London: Allen and Unwin, 1942.

Hynson, L. O., W. Kostlevy and Albert C. Outler. *The Wesleyan Revival: John Wesley's Ethics for Church and State*. Salem, OH: Schmul Publishing Company, 1999.

Ingersol, Stan, and Wesley Tracy. *Here We Stand: Where Nazarenes Fit in the Religious Marketplace*. Kansas City, MO: Beacon Hill Press, 1999.

Jablonka, E., M. J. Lamb and A. Zeligowski. *Evolution in Four Dimensions: Genetic, Epigenetic, Behavioral, and Symbolic Variation in the History of Life*. Rev. ed. Cambridge, MA: MIT Press, 2014.

Janzen, D. H. "Coevolution of Mutualism Between Ants and Acacias in Central America." *Evolution* 20 (1966): 249-75.

———. "How to Be a Fig." *Annual Review of Ecology and Systematics* 10 (1979): 13-52.

Jencks, Charles. "EP, Phone Home." In *Alas, Poor Darwin: Arguments Against Evolutionary Psychology*, edited by Hilary Rose and Steven P. R. Rose. New York: Harmony Books, 2000.

Kamin, Leon J. *The Science and Politics of I.Q.* Mahwah, NJ: Lawrence Erlbaum Associates, Inc., 1974.

Kant, Immanuel. *The Critique of Judgement*. Translated by James Creed Meredith. 2 vols. Oxford: Clarendon Press, 1952.

———. *Immanuel Kant's Critique of Pure Reason*. Translated by Norman Kemp Smith. London: Macmillan, 1929.

Kitcher, Philip. *Vaulting Ambition: Sociobiology and the Quest for Human Nature.* Cambridge, MA: MIT Press, 1985.

Lack, David Lambert. *Population Studies of Birds.* Oxford: Clarendon Press, 1966.

Lewontin, Richard C. *Biology as Ideology: The Doctrine of DNA.* New York: HarperPerennial, 1992.

Lewontin, Richard C., Leon J. Kamin and Steven P. R. Rose. *Not in Our Genes: Biology, Ideology, and Human Nature.* New York: Pantheon Books, 1984.

Long, D. Stephen. *John Wesley's Moral Theology: The Quest for God and Goodness.* Nashville: Kingswood Books, 2005.

Lowe, E. Jonathan. "Rational Selves and Freedom of Action." *Personal Agency* 1 (2008): 179-99.

Lucas, J. R. *Responsibility.* Oxford: Oxford University Press, 1993.

MacIntyre, Alasdair C. *After Virtue: A Study in Moral Theory.* 3rd ed. Notre Dame, IN: University of Notre Dame Press, 2007.

———. *Dependent Rational Animals: Why Human Beings Need the Virtues.* The Paul Carus Lecture Series. Chicago: Open Court, 1999.

Maddox, Randy L. "John Wesley and Eastern Orthodoxy: Influences, Convergences, and Differences." *Asbury Theological Journal* 45, no. 2 (1990): 29-53.

———. *Responsible Grace: John Wesley's Practical Theology.* Nashville: Kingswood Books, 1994.

Mann, Janet. "Nurturance or Negligence: Maternal Psychology and Behavioral Preference Among Preterm Twins." In *The Adapted Mind: Evolutionary Psychology and the Generation of Culture,* edited by Jerome H. Barkow, Leda Cosmides and John Tooby, 367-90. New York: Oxford University Press, 1992.

Mann, Mark H. *Perfecting Grace: Holiness, Human Being, and the Sciences.* New York: Bloomsbury Academic, 2006.

Marquardt, Manfred. *John Wesley's Social Ethics: Praxis and Principles.* Nashville: Abingdon Press, 1992.

McCormick, K. Steve. "Theosis in Chrysostom and Wesley: An Eastern Paradigm of Faith and Love." *Wesley Theological Journal* 26 (1991): 38-103.

McTyeire, Holland Nimmons. *A History of Methodism: Comprising a View of the Rise of This Revival of Spiritual Religion in the First Half of the Eighteenth Century, and of the Principal Agents by Whom It Was Promoted in Europe and America; With Some Account of the Doctrine and Polity of Episcopal Methodism in the United States, and the Means and Manner of Its Extension Down to A.D. 1884.* Nashville: Southern Methodist Publishing House, 1884.

Messer, Neil. *Selfish Genes and Christian Ethics: Theological and Ethical Reflections on Evolutionary Biology.* London: SCM, 2007.

Midgley, Mary. *The Ethical Primate: Humans, Freedom, and Morality.* London: Routledge, 1994.

———. *Evolution as a Religion: Strange Hopes and Stranger Fears.* University Paperbacks. London: Methuen, 1985.

———. "Gene-Juggling." *Philosophy* 54, no. 210 (1979): 439-58.

———. *The Myths We Live By.* Routledge Classics. New York: Routledge, 2011.

———. "Why Memes?" In *Alas, Poor Darwin: Arguments Against Evolutionary Psychology,* edited by Hilary Rose and Steven P. R. Rose. New York: Harmony Books, 2000.

Moore, Darrell. "Classical Wesleyanism." Unpublished paper, 2011.

Moore, G. E. *Principia Ethica.* Dover Philosophical Classics. Mineola, NY: Dover Publications, 2004.

Moore, Marselle. "Development in Wesley's Thought on Sanctification and Perfection." *Wesley Theological Journal* 20, no. 2 (1985): 29-53.

Morris, Simon Conway. *The Crucible of Creation: The Burgess Shale and the Rise of Animals.* Oxford: Oxford University Press, 1998.

Mühling, Markus. *Resonances: Neurobiology, Evolution and Theology; Evolutionary Niche Construction, the Ecological Brain and Relational-Narrative Theology.* Göttingen: Vandenhoeck & Ruprecht, 2014.

Munafò, Marcus R., Elaine C. Johnstone, Paul Aveyard and Theresa Marteau. "Lack of Association of Oprm1 Genotype and Smoking Cessation." *Nicotine & Tobacco Research,* September 2012.

Nelkin, Dorothy. "Less Selfish Than Sacred: Genes and the Religious Impulse in Evolutionary Psychology." In *Alas, Poor Darwin: Arguments Against Evolutionary Psychology,* edited by Hilary Rose and Steven P. R. Rose. New York: Harmony Books, 2000.

Nowak, Martin A. "Five Rules for the Evolution of Cooperation." In *Evolution, Games, and God: The Principle of Cooperation,* edited by Martin A. Nowak and Sarah Coakley. Cambridge, MA: Harvard University Press, 2013.

Nowak, Martin A., Corina E. Tarnita and Edward O. Wilson. "The Evolution of Eusociality." *Nature* 466, no. 7310 (2010).

Nurnberger, John I., and Laura Jean Bierut. "Seeking the Connections: Alcoholism and Our Genes." *Scientific American,* no. 296 (April 2007): 46-53.

Oden, Thomas C. *John Wesley's Scriptural Christianity: A Plain Exposition of His Teaching on Christian Doctrine.* Grand Rapids: Zondervan, 1994.

————. *John Wesley's Teachings: Christ and Salvation.* Vol. 2. Grand Rapids: Zondervan, 2012.

O'Hear, Anthony. *Beyond Evolution: Human Nature and the Limits of Evolutionary Explanation.* Oxford: Clarendon Press, 1997.

Olsen, Kirstin. *Daily Life in 18th-Century England.* Daily Life Through History. Westport, CT: Greenwood Press, 1999.

Oord, Thomas Jay. *The Altruism Reader: Selections from Writings on Love, Religion, and Science.* West Conshohoken, PA: Templeton Foundation Press, 2008.

————. *Defining Love: A Philosophical, Scientific, and Theological Engagement.* Grand Rapids: Brazos Press, 2010.

————. "The Love Racket: Defining Love and Agape for the Love-and-Science Research Program." In *The Altruism Reader: Selections from Writings on Love, Religion, and Science,* edited by Thomas Jay Oord. West Conshohoken, PA: Templeton Foundation Press, 2008.

————. "Morals, Love, and Relations in Evolutionary Theory." In *Evolution and Ethics: Human Morality in Biological and Religious Perspective,* edited by Philip Clayton and Jeffrey Schloss, 287-301. Grand Rapids: Eerdmans, 2004.

Origen. "Dialogue with Heraclides." In *Alexandrian Christianity,* edited by John Ernest Leonard Oulton and Henry Chadwick. The Library of Christian Classics. Philadelphia: Westminster Press, 1954.

Outka, Gene H. *Agape: An Ethical Analysis.* Yale Publications in Religion 17. New Haven, CT: Yale University Press, 1972.

Outka, Gene H., Edmund N. Santurri and William Werpehowski. *The Love Commandments: Essays in Christian Ethics and Moral Philosophy.* Washington, DC: Georgetown University Press, 1992.

Outler, Albert Cook. *Theology in the Wesleyan Spirit.* Nashville: Tidings, 1975.

Peacocke, A. R. *God and the New Biology.* San Francisco: Harper & Row, 1986.

————. *Theology for a Scientific Age: Being and Becoming—Natural, Divine, and Human.* Theology and the Sciences. Minneapolis: Fortress Press, 1993.

Piette, Maximin, and Joseph Bernard Howard. *John Wesley in the Evolution of Protestantism.* New York: Sheed & Ward, 1937.

Plotkin, Henry C. *Evolution in Mind: An Introduction to Evolutionary Psychology.* Cambridge, MA: Harvard University Press, 1998.

Pohl, Christine D. *Making Room: Recovering Hospitality as a Christian Tradition.* Grand Rapids: Eerdmans, 1999.

Pope, Stephen J. *The Evolution of Altruism and the Ordering of Love.* Moral Traditions & Moral Arguments. Washington, DC: Georgetown University Press, 1994.

————. *Human Evolution and Christian Ethics*. New Studies in Christian Ethics. Cambridge: Cambridge University Press, 2007.

————. "Relating Self, Others, and Sacrifice in the Ordering of Love." In *Altruism and Altruistic Love: Science, Philosophy, and Religion in Dialogue*, edited by Stephen G. Post, Lynn G. Underwood, Jeffrey P. Schloss and William B. Hurlbut, 168-81. New York: Oxford University Press, 2002.

————. "Varieties of Reductionism." In *Human Evolution and Christian Ethics*, 56-75. Cambridge: Cambridge University Press, 2007.

Porter, Jean. *Nature as Reason: A Thomistic Theory of the Natural Law*. Grand Rapids: Eerdmans, 2005.

Post, Stephen Garrard. *A Theory of Agape: On the Meaning of Christian Love*. Lewisburg, PA: Bucknell University Press, 1990.

————. *Unlimited Love: Altruism, Compassion, and Service*. Philadelphia: Templeton Foundation Press, 2003.

Post, Stephen G., Lynn G. Underwood, Jeffrey P. Schloss and William B. Hurlbut, eds. *Altruism and Altruistic Love: Science, Philosophy, and Religion in Dialogue*. New York: Oxford University Press, 2002.

Rachels, James. *Created from Animals: The Moral Implications of Darwinism*. Oxford: Oxford University Press, 1990.

Rack, Henry D. *Reasonable Enthusiast: John Wesley and the Rise of Methodism*. 2nd ed. Nashville: Abingdon Press, 1993.

Rand, Ayn, and Nathaniel Branden. *The Virtue of Selfishness: A New Concept of Egoism*. New York: New American Library, 1965.

Reid, Thomas. *Essays on the Powers of the Human Mind; To Which Are Added, an Essay on Quantity, and an Analysis of Aristotle's Logic*. London: T. Tegg, 1827.

————. "The Liberty of Moral Agents." In *Delight in Thinking: An Introduction to Philosophy Reader*, edited by Scott C. Lowe and Steven D. Hales. New York: McGraw-Hill, 2007.

Richerson, P. J., and R. Boyd. "Complex Societies: The Evolutionary Origins of a Crude Superorganism." *Human Nature* 10 (1999): 253-90.

Ridley, Matt. *The Origins of Virtue: Human Instincts and the Evolution of Cooperation*. New York: Viking, 1997.

————. *The Red Queen: Sex and the Evolution of Human Nature*. New York: Macmillan Publishing Company, 1994.

Rolston, Holmes. *Genes, Genesis, and God: Values and Their Origins in Natural and Human History*. Cambridge: Cambridge University Press, 1999.

————. "The Good Samaritan and His Genes." In *Evolution and Ethics: Human Morality in Biological and Religious Perspective*, edited by Philip Clayton and Jeffrey Schloss. Grand Rapids: Eerdmans, 2004.

————. *Science and Religion: A Critical Survey*. Philadelphia: Temple University Press, 1987.

Rose, Steven P. R. "Escaping Evolutionary Psychology." In *Alas, Poor Darwin: Arguments Against Evolutionary Psychology*, edited by Hilary Rose and Steven P. R. Rose. New York: Harmony Books, 2000.

————. *Lifelines: Biology Beyond Determinism*. Oxford: Oxford University Press, 1997.

Runyon, Theodore. "The New Creation: The Wesleyan Distinctive." *Wesley Theological Journal* 31, no. 2 (1996): 5-19.

Ruse, Michael. "Charles Darwin and Group Selection." *Annals of Science* 37, no. 6 (1980): 615-30.

————. "Evolutionary Ethics: A Phoenix Arisen." *Zygon* 21, no. 1 (1986): 95-112.

————. "Evolutionary Ethics Past and Present." In *Evolution and Ethics: Human Morality in Biological and Religious Perspective*, edited by Philip Clayton and Jeffrey Schloss. Grand Rapids: Eerdmans, 2004.

————. "Evolutionary Theory and Christian Ethics: Are They in Harmony?" *Zygon* 29, no. 1 (1994): 5-24.

————. *Taking Darwin Seriously: A Naturalistic Approach to Philosophy*. New York: Blackwell, 1986.

Ruse, Michael, and Edward O. Wilson. "The Evolution of Ethics." *New Scientist* 108 (October 1985).

————. "Moral Philosophy as Applied Science." *Philosophy* 61, no. 236 (1986): 173-92.

Rushton, J. Philippe, and Richard M. Sorrentino. *Altruism and Helping Behavior: Social, Personality, and Developmental Perspectives*. Hillsdale, NJ: L. Erlbaum Associates, 1981.

Sartre, J. P. *L'être et le néant: Essai d'ontologie phénoménologique*. Paris: Gallimard, 1976.

Schaik, C. P. van. "Why Are Diurnal Primates Living in Groups?" *Behaviour* 87, no. 1-2 (1983): 120-44.

Schloss, Jeffrey P. "Emerging Accounts of Altruism: 'Love Creation's Final Law'?" In *Altruism and Altruistic Love: Science, Philosophy, and Religion in Dialogue*, edited by Stephen G. Post, Lynn G. Underwood, Jeffrey P. Schloss and William B. Hurlbut, 212-42. New York: Oxford University Press, 2002.

Schmidt, Martin. *John Wesley: A Theological Biography*. New York: Abingdon Press, 1963.

Schweiker, William. *Responsibility and Christian Ethics*. New Studies in Christian Ethics. Cambridge: Cambridge University Press, 1995.

Smith, Christian, Michael O. Emerson and Patricia Snell. *Passing the Plate: Why American Christians Don't Give Away More Money*. Oxford: Oxford University Press, 2008.

Snyder, Howard A. *The Radical Wesley & Patterns for Church Renewal*. Downers Grove, IL: InterVarsity Press, 1980.

Snyder, Howard A., and Joel Scandrett. *Salvation Means Creation Healed: The Ecology of Sin and Grace; Overcoming the Divorce Between Earth and Heaven*. Eugene, OR: Cascade Books, 2011.

Sober, Elliott, and David Sloan Wilson. *Unto Others: The Evolution and Psychology of Unselfish Behavior*. Cambridge, MA: Harvard University Press, 1998.

Song, Robert. *Human Genetics: Fabricating the Future*. Ethics & Theology. London: Darton, Longman and Todd, 2002.

———. "The Human Genome Project as Soteriological Project." In *Brave New World: Theology, Ethics, and the Human Genome*, edited by Celia Deane-Drummond, 164-84. London: T&T Clark, 2003.

Southey, Robert. *The Life of Wesley: And the Rise and Progress of Methodism*. London: Frederick Warne, 1889.

Stavropoulos, Christoforos. "Partakers of Divine Nature." In *Eastern Orthodox Theology: A Contemporary Reader*, edited by D. B. Clendenin, 183-92. Grand Rapids: Baker Publishing Group, 2003.

Stevens, Abel. *The History of the Religious Movement of the Eighteenth Century: Called Methodism, Considered in Its Different Denominational Forms, and Its Relation to British and American Protestantism*. New York: Carlton & Porter, 1859.

Taylor, Charles. *Philosophy and the Human Sciences*. Philosophical Papers. Cambridge: Cambridge University Press, 1985.

———. "What's Wrong with Negative Liberty." In *Contemporary Political Philosophy: An Anthology*, edited by Robert E. Goodin and Philip Pettit. Blackwell Philosophy Anthologies. Malden, MA: Blackwell Publishing, 2006.

Taylor, Jeremy. *Selected Works*. Edited by Thomas K. Carroll. Classics of Western Spirituality. New York: Paulist Press, 1990.

Taylor, John. *A Reply to the Reverend Mr. John Wesley's Remarks on the Scripture-Doctrine of Original Sin to Which Is Added, a Short Inquiry into the Scripture-Sense of the Word Grace*. London: M. Waugh, 1767.

———. *The Scripture-Doctrine of Original Sin: Proposed to Free and Candid Examination in Three Parts*. London: J. Wilson, 1740.

Titus, Craig Steven. *The Psychology of Character and Virtue*. Arlington, VA: Institute for the Psychological Sciences Press, 2009.

Trivers, Robert L. "The Evolution of Reciprocal Altruism." *The Quarterly Review of Biology* 46, no. 1 (1971): 35-57.

———. "Parental Investment and Sexual Selection." In *Sexual Selection and the Descent of Man, 1871–1971*, edited by Bernard Grant Campbell, 136-79. Chicago: Aldine Pub. Co., 1972.

Troeltsch, Ernst. *The Social Teaching of the Christian Churches*. Translated by Olive Wyon. Chicago: University of Chicago Press, 1981.

———. *Die Soziallehre der Christlichen Kirchen und Gruppen, Gesammelte Schriften*. Tübingen: J. C. B. Mohr, 1912.

Tyerman, L. *The Life and Times of the Rev. John Wesley: Founder of the Methodists*. Vol. 2. London: Hodder and Stoughton, 1870.

Tyndale, Rachel F. "Genetics of Alcohol and Tobacco Use in Humans." *Annals of Medicine* 35, no. 2 (2003): 94-121.

Van Till, Howard J. *Portraits of Creation: Biblical and Scientific Perspectives on the World's Formation*. Grand Rapids: Eerdmans, 1990.

Wake, David B., and Gerhard Roth. *Complex Organismal Functions: Integration and Evolution in Vertebrates; Report of the Dahlem Workshop on Complex Organismal Functions—Integration and Evolution in Vertebrates, Berlin 1988, August 28–September 2*. Dahlem Workshop Reports. Chichester, UK: Wiley, 1989.

Wake, David B., Gerhard Roth and Marvalee H. Wake. "On the Problem of Stasis in Organismal Evolution." *Journal of Theoretical Biology* 101, no. 2 (1983): 211-24.

Watson, David Lowes. *The Early Methodist Class Meeting: Its Origins and Significance*. Nashville: Discipleship Resources, 1985.

Weiss, Kenneth M., and Anne V. Buchanan. *The Mermaid's Tale: Four Billion Years of Cooperation in the Making of Living Things*. Cambridge, MA: Harvard University Press, 2009.

Wesley, John. *The Doctrine of Original Sin: According to Scripture, Reason, and Experience*. Bristol: F. Farley, 1757.

———. *Explanatory Notes upon the New Testament*. Vol. 2, *Romans to Revelation*. Kansas City, MO: Beacon Hill Press, 1981.

———. *John Wesley: A Representative Collection of His Writings*. Edited by Albert

C. Outler. A Library of Protestant Thought. New York: Oxford University Press, 1964.

————. *Journal and Diaries*. Edited by Richard P. Heitzenrater and W. Reginald Ward. The Bicentennial Edition of the Works of John Wesley. Nashville: Abingdon Press, 1990.

————. *The Journal of the Rev. John Wesley*. Edited by N. Curnock. Whitefish, MT: Kessinger Publishing, 2006.

————. *The Letters of the Rev. John Wesley*. Edited by John Telford. Standard ed. 8 vols. London: Epworth Press, 1931.

————. *The Methodist Societies: History, Nature, and Design*. Edited by Rupert E. Davies. The Bicentennial Edition of the Works of John Wesley. Nashville: Abingdon Press, 1989.

————. *A Plain Account of the People Called Methodists in a Letter to the Revd. Mr. Perronet*. 2nd ed. London: W. Strahan, 1749.

————. *Sermons 1–33*. Edited by Albert C. Outler. The Bicentennial Edition of the Works of John Wesley. Nashville: Abingdon Press, 1984.

————. *Sermons 34–70*. Edited by Albert C. Outler. The Bicentennial Edition of the Works of John Wesley. Nashville: Abingdon Press, 1985.

————. *Sermons 71–114*. Edited by Albert C. Outler. The Bicentennial Edition of the Works of John Wesley. Nashville: Abingdon Press, 1986.

————. *Sermons 115–151*. Edited by Albert C. Outler. The Bicentennial Edition of the Works of John Wesley. Nashville: Abingdon Press, 1987.

————. *The Works of John Wesley*. Edited by Thomas Jackson and Albert C. Outler. 14 vols. Grand Rapids: Zondervan, 1958.

Wesley, John, and Charles Wesley. *Hymns and Sacred Poems*. 3rd ed. London: W. Strahan, 1739.

Wiebes, J. T. "A Short History of Fig Wasp Research." *Gardens Bulletin,* 1976, 207-32.

Williams, George C. *The Pony Fish's Glow: And Other Clues to Plan and Purpose in Nature*. Science Masters Series. New York: BasicBooks, 1997.

Wilson, David Sloan. *Does Altruism Exist? Culture, Genes, and the Welfare of Others*. Foundational Questions in Science. New Haven, CT: Yale University Press and Templeton Press, 2015.

Wilson, Edward O. *Consilience: The Unity of Knowledge*. New York: Random House, 1998.

————. *On Human Nature*. Cambridge, MA: Harvard University Press, 1978.

————. *On Human Nature*. 25th anniversary ed. Cambridge, MA: Harvard University Press, 2004.

———. *The Social Conquest of Earth*. New York: Liveright Pub. Corporation, 2012.

———. *Sociobiology: The New Synthesis*. Cambridge, MA: Belknap Press of Harvard University Press, 1975.

Witherington, Ben, and Darlene Hyatt. *Paul's Letter to the Romans: A Socio-Rhetorical Commentary*. Grand Rapids: Eerdmans, 2004.

Wrangham, Richard W. "An Ecological Model of Female-Bonded Primate Groups." *Behaviour* 75, nos. 3–4 (1980): 262-300.

Wrangham, Richard W., and Dale Peterson. *Demonic Males: Apes and the Origins of Human Violence*. Boston: Houghton Mifflin, 1996.

Wright, Robert. *The Moral Animal: The New Science of Evolutionary Psychology*. New York: Pantheon Books, 1994.

Wuthnow, Robert. *God and Mammon in America*. Toronto: Free Press, 1994.

Wynkoop, Mildred Bangs. *A Theology of Love: The Dynamic of Wesleyanism*. Kansas City, MO: Beacon Hill Press, 1972.

Wynne-Edwards, Vero Copner. *Animal Dispersion in Relation to Social Behaviour*. New York: Hafner Pub. Co., 1962.

Young, T. Kue. *The Health of Native Americans: Toward a Biocultural Epidemiology*. New York: Oxford University Press, 1994.

Zhang, Y., D. Wang, A. D. Johnson, A. C. Papp and W. Sadée. "Allelic Expression Imbalance of Human Mu Opioid Receptor (Oprm1) Caused by Variant A118g." *The Journal of Biological Chemistry* 280, no. 38 (2005): 32618-24.

GENERAL INDEX

à Kempis, Thomas, 148
Abraham, William J., 139-40
Adam, 121, 139, 152, 169, 212-13
addiction, 118, 123, 160
agency, 25, 30, 34, 38, 42, 44,
48, 49, 50, 51, 52, 60-61, 63,
73, 75, 79-80, 83, 85, 87, 91,
96, 99, 101, 105, 108, 116,
118-23, 125, 127, 130, 135-36,
139, 163, 165, 183
alcoholism, 17, 118, 123, 174,
197
Alexander, Richard D., 44, 46,
48, 50, 56, 79
altruism, 18-39, 41-57, 59-69,
71-89, 91, 93-107, 109, 111,
113-17, 119-21, 123, 125, 127,
129, 131, 133, 135, 139-40,
172-75, 177, 179, 181, 183,
185-87, 189, 191, 192-93, 195,
197-99, 201-6, 208, 210-11,
213-14
    kin, 22, 25, 43-45, 51, 54,
    73, 91, 204
    reciprocal, 22, 25, 42, 47,
    49-50, 66
animals, 11, 23, 24-26, 30, 35,
36, 38, 40, 42, 45-46, 51, 57,
59, 67-68, 74, 75-76, 77, 81,
83, 86, 88, 93, 96, 103, 104,
107, 109, 112, 113, 199, 125,
126-30, 134-35, 145, 202
anthropology,
anthropological, 35, 38, 74,
75, 127, 146, 163, 165, 167
ants, 45, 48, 99-100
Aquinas, Thomas, 36, 67, 100,
101
    Thomism, 36, 41-42, 58,
    67-68, 91, 98, 101, 215
Aristotle, Aristotelian, 21, 36,
47, 68, 81, 120, 128, 130
Arnhart, Larry, 67, 84

Augustine, Augustinian, 125,
145, 147, 158, 169, 190, 197,
206, 212-15
Axelrod, Robert M., 40, 44,
46-51, 85
Ayala, Francisco, 114
bands, band meetings, 11-12,
18-20, 28, 140, 153, 156, 158,
174, 178-82, 186-89, 191-92,
201, 203, 206, 211, 213, 216,
224-26
Barbour, Ian G., 214-15
Batson, C. Daniel, 34, 61, 101
behavior
    animal, 23, 25-26, 38, 42,
    46, 51, 57, 67-68, 75, 77,
    81, 83, 86, 88, 93, 96, 103,
    104, 107, 109, 112, 113, 119,
    125-28, 134-35, 202, 234
    human, 18-20, 22-23,
    25-29, 32, 34, 38, 41-42,
    44, 46-47, 50-51, 53-65,
    67-69, 72-73, 75, 77-89,
    91-99, 102-10, 112-16,
    118-20, 123-28, 131-35,
    137-38, 144, 165, 172-74,
    197, 201-5, 215-16
Berry, Wendell, 89, 100
biocultural epidemiology, 17
biology, 20, 22, 24-26, 35, 38,
44, 53, 57, 63-67, 73-74, 78, 81,
84-85, 94, 106, 108, 110-14, 117,
123, 126, 130, 132-34, 137-38,
144, 161, 168-70, 172, 202
Bowker, John, 64-65, 73-74,
103-4
Calvin, Calvinists, 125, 147-48,
206
character, 21, 28, 30, 70, 85, 99,
107, 125, 130, 146, 159, 166,
174, 194, 207, 211
Chilcote, Paul Wesley, 30, 142,
178, 181, 186, 193, 199, 205

choice, 21, 27, 36, 50-51, 68, 74,
99, 107-8, 110, 112, 114, 117-18,
120, 122-23, 125, 128-31, 135,
146, 161-62, 198, 202
Christensen, Michael J., 138,
151, 153, 166
Christian ethics, 23, 31-35, 78,
92, 95-101, 129, 197-98, 202,
210
Christian perfection, 20-21,
27-28, 137-41, 144, 148-56,
158-71, 189, 193, 201, 203-4,
206-7, 212-13, 216, 218-19,
221, 223
church, 31-32, 36, 141-43, 160,
165, 175-83, 185-90, 193-95,
199, 208, 210, 211, 216-17, 226
Churchland, Patricia S., 108,
110-12, 160, 161
Clark, Stephen R. L., 38, 73,
82, 108, 112, 117
class meeting, 177, 178, 183
Classical Wesleyanism, 29,
142-43, 173-74, 177, 188,
190-91
Clayton, Philip, 26, 35-36, 38,
41, 47, 55, 61, 65, 67, 72, 103,
215
Coakley, Sarah, 37-38, 202
coevolution, 48, 65, 72
Collins, Kenneth J., 156, 158,
179, 182, 187, 202
common grace, 206
community, 20, 22, 27, 30-31,
36, 40, 43, 65, 80-81, 88, 101,
141, 161, 173-74, 181, 185-86,
194, 197-98, 203, 205, 207,
211, 216
compatibilism, 117
consciousness, 26, 47, 108, 110,
111-12, 114, 123, 126, 129-30
consilience, 56, 79, 89, 116-17,
130

constraint, 20, 25-30, 33, 36,
    60, 61, 64, 84, 94, 104, 106-9,
    114, 119-28, 131, 135, 138-39,
    142, 144, 157-59, 163, 168-73,
    184, 186, 187, 189-92, 196-97,
    199, 200, 203-7, 211, 215, 216
cooperation, 12, 24, 37, 40-42,
    44-51, 60, 66, 82, 85, 97, 114,
    129, 133, 202, 207, 211, 215-16
creation, 23, 30, 33, 35-36, 87,
    90, 92, 107, 113, 129, 136,
    144-45, 150, 152, 154, 159-62,
    164, 171, 200, 212-13, 215
creationism, 42, 65, 116
Crick, Francis, 109
culpability, 120, 122-23, 125, 129
culture, 22-23, 25-26, 29, 45,
    54, 63-65, 68-69, 71, 88, 95,
    98, 111, 116, 126, 132, 142, 172,
    204, 214
Cunningham, Conor, 42, 54,
    65, 93, 96, 116-17
Darwin, Charles, 10, 26, 32,
    38-40, 42-44, 47, 53-55, 58,
    61, 65-71, 76, 83-85, 87, 93,
    96, 100, 109-10, 116-17, 125,
    127-28, 134, 174, 204
Darwinism, 54, 61, 96, 109, 128
Dawkins, Richard, 23-25, 32,
    39, 41-42, 44, 55-58, 65, 67,
    69-76, 78-82, 86, 89-92, 96,
    100, 104, 105, 108, 110,
    113-16, 119, 125, 127-28,
    133-35, 205
De Waal, Frans, 20, 24, 29, 32,
    59, 65, 81-84, 92-93, 139, 198
Deane-Drummond, Celia, 19,
    21, 22, 24, 113
Dennett, Daniel C., 25-26, 38,
    50, 69, 108, 110, 112-14, 117,
    122-23, 125-27, 130-31, 134-35,
    139
denomination, 161, 191, 210
dependency, 30, 43, 46, 111,
    120-21, 126, 143, 170, 193, 197
Descartes, René, 96, 123, 126,
    134
Descent of Man, The, 43, 55,
    84-85, 128
determinism, 23, 62, 85, 107-8,
    111, 113, 115, 117, 119, 210
dichotomy, 85, 89, 92, 115, 176

discipleship, 30, 146, 152, 178,
    183, 193, 199
doctrine, 10, 22, 29, 36, 53, 63,
    138-39, 141-42, 144-48,
    150-53, 156, 160, 169-70, 173,
    177, 183, 199, 205, 222
Dover, Gabriel, 70-72
Drees, Willem B., 53, 84, 90,
    92, 94-95, 97
dualism, 27, 36, 108, 110,
    115-16, 123, 126, 134, 142-43,
    159, 162-65, 170
Dunnington, Kent, 118, 123
Eagleman, David, 122-23
Eastern Orthodoxy, 19, 141,
    145, 147, 149, 150-53, 166,
    168-69, 208-9, 212-15
egalitarian, 34
emotions, 52, 112, 129, 131
empathy, 24, 47, 82-84
entire sanctification, 138, 149,
    162-63, 171
environment, environmental,
    17-18, 20, 22-23, 25-30, 33, 35,
    37-41, 43-44, 47-51, 54-58,
    60, 62-67, 69-70, 72, 76,
    84-87, 89, 97, 99, 100, 102,
    104-7, 109, 111, 113, 116-17,
    119-26, 132-35, 139, 172, 186,
    196, 198, 200-201, 203-4,
    206-7, 210-11, 215
epigenetics, 116, 135
Episcopal, 183
ethics, 18, 20-36, 38, 41, 43, 45,
    47, 50, 54, 55, 59-61, 65-67,
    72-82, 84-85, 87, 89, 92,
    94-101, 104-5, 108-10, 112,
    114-15, 117, 119-35, 137, 142, 144,
    153, 158, 165, 179, 182, 197-98,
    201-5, 210, 212-13, 215-16
eusociality, 52
evolution, 18-19, 21-27, 30-63,
    65-72, 74-114, 116, 119-20,
    122, 125-139, 144, 158-59,
    168-69, 177, 197-98, 201-5,
    208, 210, 214-16
evolutionary biology, 10, 12,
    18, 35, 69, 78, 83, 96, 132,
    137-38, 144, 168-69, 201, 215
evolutionary psychology, 41,
    42, 58, 67-68, 70, 91, 112, 132,
    205, 215

exchange, 82
explanation, 18-23, 27, 31-32,
    34-35, 37, 52, 57, 59, 62-65,
    67-73, 75-77, 79, 81, 83-85, 87,
    91-92, 95-98, 102-5, 120, 136,
    151, 161, 166, 201-2, 204-5, 215
faith, 18, 21, 30-31, 33, 43, 82,
    94-95, 101, 141-44, 147,
    149-52, 154, 156, 159, 177-79,
    189, 191-93, 195, 205, 207,
    208, 216, 218-19, 222, 225
fall, fallenness, 61, 117, 125, 131,
    139, 142, 145, 159-161, 166,
    168-70, 205-7, 212-13, 215
fallacy, 22, 51, 64, 87, 96-98,
    102-3
false dilemma, false
    opposition, 22, 64, 89, 96,
    102
Fodor, Jerry A., 69
forlorn existence, 59
free will, 18, 26-27, 35, 64,
    86-87, 104-5, 107-11, 115-17,
    120, 122, 125-26, 132-34, 158,
    162, 167, 169-70, 197, 201,
    204, 215-16
freedom, 25-27, 30, 38, 43, 50,
    53, 60, 73-74, 76, 97, 104,
    106-10, 114, 117-31, 133-35,
    145, 151, 155, 158, 162-63,
    167-70, 172, 186, 189-90,
    205-7, 218, 224
game theory, 22, 47, 49-51,
    204
gender, 46, 179
generosity, 19, 24, 75, 79, 96,
    140, 198, 203, 210, 211
genetics, 17-35, 37-39, 41-93,
    97, 99-100, 102-7, 109, 111-23,
    125, 127-28, 130-35, 144,
    158-63, 166-68, 170, 172, 174,
    197, 201-3, 205, 207, 212-15
God, 18, 21, 25, 29, 32-33, 35,
    37-38, 58, 66, 75, 78, 92, 97,
    100-101, 107, 122, 125-26,
    129-31, 134, 136, 138-44,
    146-71, 173, 182-90, 192-94,
    196-200, 202, 205-7, 209-12,
    214-21, 224-26
good natured, 59
Gould, Stephen Jay, 39-40, 57,
    69, 95

grace, 28-29, 61, 76, 100, 125, 138-43, 145-54, 156-63, 166-71, 173, 180, 182, 186-87, 189, 192, 194, 196-98, 204-7, 209-10, 214-17, 222
Gundry, Stanley N., 209
Gunton, Colin E., 144
Hamilton, W. D., 46, 50, 93
Hauerwas, Stanley, 20, 203
Henderson, D. Michael, 183
Hick, John, 214
Hobbes, Thomas, 47
holiness, 10, 12, 14, 18-21, 27-31, 33, 94, 137-218
    Wesleyan, 18-19, 27, 137, 139, 141, 143, 145, 147, 149, 151, 153, 155, 157, 159, 161, 163, 165, 167, 169, 171, 208, 210
Hölldobler, Bert, 79
Homo sapiens, 27, 38, 39
honesty, 80
Hudson, W. D., 98
human genome, 22, 55
human nature, 19-20, 29, 51, 54, 56, 58, 61, 67, 76, 78-79, 87, 91, 95, 98, 100, 102-4, 108, 112, 129, 130, 134, 139, 142-48, 157, 173, 192, 198, 203-4, 207, 210, 214
humans, 18-20, 22-36, 41-42, 50-69, 71-84, 87-108, 110-52, 156-59, 161-66, 168-74, 187, 192, 197-98, 200-207, 210, 213-16
Huxley, Julian, 54, 92
hymns, 193, 200
ideology, 53, 73, 95
intercline ordering, 104
Irenaeus, 209
is-ought, 96, 98, 100, 121
Janzen, D. H., 48
Jesus Christ, 18, 85, 102, 149, 155, 161, 200, 212, 219-20, 225
journal (Wesley), 143, 146, 156-58, 174, 185, 187, 188
Kant, Immanuel, 131
kin, 22, 25, 34, 38, 42-48, 51-52, 54, 58, 67, 73, 81-82, 84-86, 98, 103, 204
kin altruism. *See* altruism, kin
kindness, 24, 35, 219
Kitcher, Philip, 78

letters (Wesley), 143, 147, 206
Lewontin, Richard C., 22, 53, 63, 65, 88, 112, 115
Long, D. Stephen, 182
love, 18, 30-32, 36, 38, 43, 56, 59-60, 74, 78, 97-98, 100-102, 112, 138, 140-41, 149-52, 154, 155, 157, 167-69, 171, 174, 177, 180, 186-88, 190-95, 197-98, 205-8, 217, 219-22, 225-26
MacIntyre, Alasdair C., 21, 30, 216
Mann, Mark H., 204
Marquardt, Manfred, 179
McCormick, K. Steve, 208
meaning, 18, 36, 92, 151
meme, 54, 58, 64-65, 67-69, 71-73, 86, 89, 113-14, 116, 125
Messer, Neil, 18, 24-25, 28, 30-35, 78, 84, 131-33, 143-44, 202
metaphysics, 58, 90, 91, 111, 124, 205
method, 36, 78, 94, 147, 185, 194
Methodism, 20, 27-29, 140, 142, 143, 150, 155, 157-58, 169-70, 172, 174-87, 191-92, 194-97, 199, 203
Midgley, Mary, 68-69, 71, 78, 81-82, 89-91, 103, 130
minds, 25, 77, 104, 107, 110, 112-14, 121-23, 126, 140-41, 155-56, 171, 210
miracles, 89, 188
modern science, 69, 111, 144
modern synthesis, 43, 54
Moore, Darrell, 29, 142-43, 173-74
Moore, G. E., 98
Moore, Marselle, 138, 141, 147-48, 155, 171
moral philosophy, 66, 98
moral theology, 64, 94-95, 102
morality, 18, 20-22, 25-28, 30-35, 43, 51-52, 54-59, 64-67, 72, 74-75, 79, 81, 87, 91-92, 94-99, 101-3, 106-8, 110-11, 119-30, 133-34, 141, 144, 145, 158, 161, 174, 195, 197-98, 201-3, 208, 211, 215
    human, 26, 35, 38, 65, 215
Morris, Simon Conway, 23, 95

Mühling, Markus, 108, 214
mutation, 39, 70, 72, 82
natural selection, 10, 26, 38, 39, 42, 48, 54-57, 61-62, 69-71, 78, 84, 93, 107, 111-12, 114, 119, 132
naturalism, 97, 126
nature, natural, 24, 30, 36, 41, 52-59, 61, 65-68, 71, 75-76, 91, 93, 122, 127, 128, 130, 136, 140
nature/nurture, 57, 59 65, 66, 87, 139
neurons, 47, 109
neuroscience, 25, 111, 122, 164
New Biology, 113, 134
Nicomachean Ethics, 47, 128, 130
Nowak, Martin A., 37-38, 42, 52
nurture, 28-29, 139, 173
obligation, 43, 81, 98
Oden, Thomas C., 146, 205-7, 215-16
O'Hear, Anthony, 76
Oord, Thomas Jay, 30, 32, 35-36, 41, 45, 50, 53, 60, 107
ordering, 61, 98, 100-102
Origen, 153, 209
Origin of Species, The, 39, 53, 96
original sin, 28-29, 93-94, 96, 137, 139, 143-47, 158-59, 162, 167, 169-71, 173, 206-7, 213
Outka, Gene H., 101
Outler, Albert Cook, 29, 138, 140-45, 147-54, 168, 173, 175, 192-96, 205-6
overcoming, 17, 25-28, 105-7, 119-21, 125, 127-28, 130-31, 134-35, 157, 159, 161-63, 166-71, 192, 196, 207, 210, 212, 225
patristics, 138, 151
Peacocke, A. R., 108, 134
perfection, 20-21, 27-28, 71, 137-44, 148-56, 158-63, 165-71, 189, 193, 201, 203-4, 206-7, 212-13, 216, 218-19, 221-23. *See also* Christian perfection
philosophy, 43, 47, 86, 94, 96, 102, 108, 214

Plato, Platonism, 128, 142
Pohl, Christine D., 173
politics, 88, 142
Pope, Stephen J., 23, 25-26,
  30-35, 54, 57, 60, 65-66, 74,
  79-80, 85, 87, 91, 95, 99, 114,
  119, 127, 129-31, 134, 136,
  197-98, 202, 204, 208, 210,
  216
Porter, Jean, 36
Post, Stephen G., 44, 87, 98,
  101
predation, 88
prevenient grace, 28-29,
  139-40, 142-43, 145-47, 150,
  157-58, 162, 167, 170, 173, 187,
  192, 197, 206-7, 214
primates, 24, 32, 49, 65, 92-93,
  103, 104
Protestantism, 32, 141, 143-44,
  148, 150, 153, 209
psychology, 23, 30, 33-34, 37,
  59, 61, 70, 73, 101, 112
punctuated equilibrium, 40,
  57
Rachels, James, 128
Rack, Henry D., 177
Rand, Ayn, 51, 128
reasonable enthusiast, 177
reciprocal altruism. See
  altruism, reciprocal
red queen, 51, 90, 99, 109, 114
reductionism, 22-25, 27-28, 33,
  35, 58, 62-63, 70, 76-79,
  81-82, 84, 87, 89, 97-98, 102,
  104-5, 108, 112, 115, 146, 164,
  205
Reformed, 19, 32, 143, 148,
  209-10
Reid, Thomas, 120
Religion, 35-36, 53, 89-90, 92,
  94-97, 138, 147, 154, 168, 176,
  179, 183, 188, 192-93, 206, 211,
  214, 216
replication, 37-42, 44, 46,
  50-52, 55-56, 60, 70, 76, 86,
  91, 115-17
responsibility, 25-26, 31, 35, 41,
  59, 92-95, 106, 108, 115,
  120-26, 129, 134-35, 146, 163,
  166, 168, 172, 182, 190, 198,
  209-10

Ridley, Mark, 56, 93, 128
Ridley, Matt, 45-46, 51, 56, 90,
  106, 109, 113-14, 128, 133
Rolston, Holmes, 15, 55, 61, 96,
  99, 114
Roman Catholicism, 31, 36, 208
Rose, Steven P. R., 58, 68-70,
  85, 100, 112, 115
Runyon, Theodore, 150, 152,
  154
Ruse, Michael, 43, 47, 54, 59,
  68, 72, 78, 99, 100, 128, 197
sacrifice, 30, 34-35, 38, 40, 41,
  43, 45, 51-52, 54, 61, 73, 80,
  83, 86-88, 110, 155, 184, 200,
  205, 220
salvation, 29, 138, 140-45,
  148-49, 154, 156, 161, 163,
  165-70, 173, 175, 183, 198-200,
  207, 219, 221
sanctification, 150, 162
Sartre, J. P., 124
Schloss, Jeffrey P., 26, 30,
  35-36, 38, 41, 47, 55, 65, 67,
  72, 87, 215
Schweiker, William, 121-25, 129
science, 18, 22, 25, 32-33, 36,
  40, 43, 53, 58, 64, 68, 72,
  78-79, 84, 89-90, 92, 94, 96,
  97, 102, 104, 107, 111, 114, 145,
  158-59, 164, 192, 194, 201, 214
secularism, 28, 82, 122
selection, 21, 26, 34, 38-43,
  46-48, 52-57, 61-62, 69-71, 78,
  80-81, 84, 92-93, 106-7,
  109-12, 114, 119, 204, 215
selfish gene, 38, 73, 158
selfishness, 38, 56-58, 66, 73,
  89, 211
selflessness, 34-35, 43, 52, 61,
  80, 90, 110
sermon (Wesley), 138, 140-41,
  143-47, 149-55, 166, 168,
  169-70, 173, 179-80, 182, 184,
  186, 189-90, 192-96, 200,
  205, 207, 209, 213
sexuality, 25, 31, 51, 74, 81, 91,
  103, 110, 112-13, 197
sin, 28, 93-94, 96, 125, 137, 139,
  140-49, 152, 154, 156, 158-59,
  162, 164-65, 167, 169, 170-71,
  173-75, 191, 194, 196, 197,

  205-7, 209-10, 212-13, 219,
  221-22, 225-26
Smith, Christian, 31, 211
Snyder, Howard A., 159, 161,
  175, 178-81, 185-87, 189-90,
  195
Sober, Elliott, 53, 60, 61-62,
  79-80, 93, 125
social, 18-19, 22, 27, 29, 32,
  34-35, 43, 45-46, 48, 52-53,
  58, 60-61, 65, 68, 81-82,
  84-87, 93, 97-98, 101-3, 128,
  132, 145, 153-55, 176-79,
  181-82, 189, 193, 199, 202-3,
  205-6, 211
social-psychological, 34, 61, 101
society, 24-26, 31, 42, 44, 47,
  53, 56, 75, 77, 82, 88, 112, 120,
  175-76, 180-82, 184-85,
  187-88, 191, 196
sociology, sociobiology, 18-23,
  25, 27, 28, 38, 42, 44, 50,
  52-53, 57, 61-62, 64, 69-73,
  77-89, 95-96, 99-100, 102-4,
  115, 133, 137, 143, 163-66, 169,
  201-5, 212, 214-15
Song, Robert, 13, 22, 55
soul, 29, 58, 104, 110, 134, 138,
  140-42, 145, 147, 154-55, 157,
  159-66, 169-70, 173, 176, 180,
  183-84, 189, 194, 200, 218-19,
  224
Summa Theologica, 100-101
superorganism, 34, 79
superstition, 89, 92
suprascientific, 64, 73, 77-78
  89-94, 97, 102, 111, 126, 160
Taylor, Charles, 121
Taylor, John, 145-49, 155
theology, 18, 20-22, 28-29,
  32-36, 64, 76, 78, 89, 92,
  94-97, 101-2, 125, 136-37,
  139-44, 149-51, 156, 158,
  161-62, 165, 169, 171-73,
  182-83, 187, 189, 201, 204,
  206, 208, 210, 213-15
theosis, 28, 149-52, 163, 165-66,
  208-9, 212-13, 215
Thomism. See under Aquinas
Titus, Craig Steven, 211
tobacco, 118, 226
tradition, 19, 25, 32-33, 37, 65,

87, 97, 125, 133, 141, 143-45, 148-51, 161, 168, 174, 176, 208-10, 212, 215
transcendent, 32, 214
Trivers, Robert L., 66, 132
Troeltsch, Ernst, 176
truth, 33, 56, 79, 111, 140, 142, 153-54, 162, 169, 176, 194
ultra-Darwinism, 42, 65, 69, 109, 116
unconscious, 41, 48, 91, 102, 112, 122-24
unselfish, 21, 24, 66, 75, 80, 97
violence, 42, 93, 132, 157, 161, 212
virtue, 21, 29, 43, 51, 68, 99-100, 123, 129, 142, 145,

154, 166, 198, 202, 205, 210-11, 216
Wesley, Charles, 145-46, 157, 179
Wesley, John, 19-21, 27, 31, 33, 36, 94, 122, 135-59, 163-209, 213-16, 218, 221, 224
Wesleyan ethics, 21, 27-28, 35, 89, 94, 137, 158, 201, 204, 212, 215
Wesleyan spirit, 141-45, 147-48, 195-96
Wesleyanism, 18-21, 27-29, 31, 35, 89, 94, 137-39, 141, 148, 156, 158, 162-63, 168, 170, 175, 183, 197, 201, 203-4, 207-18
Western, 151, 209, 213

whole person, 62, 87, 108, 126, 132-33, 142
Williams, George C., 74-75, 84
Wilson, David Sloan, 53, 56-57, 60-61, 66-68, 79-81, 87, 89, 214
Wilson, Edward O., 25, 35, 42, 50, 52, 73, 78, 100, 103-4, 108, 117, 120, 130
works, 20, 27, 32, 126, 144, 150-56, 159, 162-63, 166-71, 173-77, 180, 189, 191-93, 196, 203, 208, 215
Wright, Robert, 112, 119, 128
Wuthnow, Robert, 211
Wynkoop, Mildred Bangs, 31, 141-42, 174, 187, 190, 216-17

STRATEGIC INITIATIVES IN EVANGELICAL THEOLOGY

IVP Academic presents a series of seminal works of scholarship with significant relevance for both evangelical scholarship and the church. Strategic Initiatives in Evangelical Theology (SIET) aims to foster interaction within the broader evangelical community and advance discussion in the wider academy around emerging, current, groundbreaking or controversial topics. The series provides a unique publishing venue for both more senior and younger promising scholars.

While SIET volumes demonstrate a depth of appreciation for evangelical theology and the current challenges and issues facing it, the series will welcome books that engage the full range of academic disciplines from theology and biblical studies, to history, literature, philosophy, the natural and social sciences, and the arts.

Editorial Advisory Board

Published Volumes

Addiction and Virtue, Kent Dunnington

The Analogy of Faith, Archie J. Spencer

Crossover Preaching, Jared E. Alcántara

The God of the Gospel, Scott R. Swain

Incarnational Humanism, Jens Zimmerman

Rethinking the Trinity & Religious Pluralism, Keith E. Johnson

Theology's Epistemological Dilemma, Kevin Diller

The Triumph of God Over Evil, William Hasker

Finding the Textbook You Need

The IVP Academic Textbook Selector
is an online tool for instantly finding the IVP books
suitable for over 250 courses across 24 disciplines.

www.ivpress.com/academic/